Classics of Science, Volume VII

Classics in Coordination Chemistry

PART II:

SELECTED PAPERS
(1798–1899)

Translated, edited, and with commentary by
GEORGE B. KAUFFMAN
Professor of Chemistry
California State College at Fresno

DOVER PUBLICATIONS, INC.
NEW YORK

For
my three girls
LAURIE, RUTH, and JUDITH

"Die Geschichte der Wissenschaft ist die Wissenschaft selbst."
(The history of science is science itself.)
—Johann Wolfgang von Goethe

Published in Canada by General Publishing Company, Ltd., 30 Lesmill Road, Don Mills, Toronto, Ontario.

Published in the United Kingdom by Constable and Company, Ltd., 10 Orange Street, London WC 2.

Classics in Coordination Chemistry, Part II: Selected Papers, 1798–1899 (Volume VII in the Dover "Classics of Science" series) contains six papers published for the first time in collected form by Dover Publications, Inc., in 1976. Five of these papers were newly translated from the German by George B. Kauffman. The translator has also provided a Preface, a General Introduction, and an Introduction and notes to each of these papers.

International Standard Book Number: 0-486-63343-8
Library of Congress Catalog Card Number: 67-26870

Manufactured in the United States of America
Dover Publications, Inc.
180 Varick Street
New York, N.Y. 10014

CLASSICS OF SCIENCE SERIES

under the General Editorship of

Gerald Holton

Professor of Physics, Harvard University

CLASSICS IN THE THEORY OF CHEMICAL COMBINATION
edited by O. Theodor Benfey

THE DISCOVERY OF RADIOACTIVITY AND TRANSMUTATION
edited by Alfred Romer

THE DEVELOPMENT OF HIGH-ENERGY ACCELERATORS
edited by M. Stanley Livingston

CLASSICS IN COORDINATION CHEMISTRY, PART I
edited by George B. Kauffman

SOURCES OF QUANTUM MECHANICS
edited by B. L. van der Waerden

RADIOCHEMISTRY AND THE DISCOVERY OF ISOTOPES
edited by Alfred Romer

CLASSICS IN COORDINATION CHEMISTRY, PART II
edited by George B. Kauffman

General Editor's Preface

This volume is one of the many planned for the series entitled *Classics of Science*. Each volume is a collection of fundamental essays and other basic original articles in a certain field of science, presented in the sequence of its development, together with an introduction, commentary, and clarifying notes by the scholar responsible for the selection of the papers. All areas of science are to be included as the series grows. Thus, in time, we shall have here a convenient network of roads to take us to the often inaccessible sources of the great rivers of science, through its widely visible mountain ranges, its renowned battlefields, and perhaps its little-known but nevertheless choice vineyards.

The articles follow the original texts verbatim, and (with rare and clearly marked exceptions) are printed in full rather than in excerpt or in edited and abbreviated versions; they may therefore be used in lieu of the original publications. Foreign-language articles are carefully translated. In short, both the interested layman and the historian or scientist should feel assured that he is reading these documents as they were intended to be read, and in the version in which they made their original contribution.

Some of these aims are similar to those of a distinguished German series of republications, begun in 1889 under the direction of the chemist and philosopher, Wilhelm Ostwald, and reaching some 400 titles before it was discontinued. There are, however, important distinctions between these two series. Ostwald's series was composed of republications of single essays, whereas here each book sets out several essays, each in the context of the others, in order to trace the development of the whole field. Hence there will be more connective commentary in this series, and the commentary can now benefit from advances made in the last three-quarters of a century, both in science and in historical scholarship.

Still, many of the aims cited in the original announcement of Ostwald's *Klassiker der exakten Wissenschaften* are again applicable:

The great progress which the sciences have made in our time, as is generally acknowledged, is in good measure owing to the cultivation and wide application of teaching methods such as demonstration lectures and laboratories. These arrangements are indeed very successful in imparting knowledge of the present state of science. But some of the most eminent persons with widest vision have frequently felt compelled to point out a deficiency that mars all too often our contemporary scientific education. This is the absence of a sense of history, and the lack of knowledge concerning those great contributions on which the edifice of science rests.

Although few instructors in scientific subjects would fail to make reference to those foundations when the occasion arises, such references remain, nevertheless, generally ineffective because the source materials of science are rarely accessible. They can be obtained only in the large libraries, and then only in single copy, so that the student is all too easily discouraged from pursuing the lead.

This lack is to be remedied through the publication of the *Klassiker der exakten Wissenschaften.* In convenient format and at reasonable prices, the seminal publications of all the exact sciences will be made available to instructors and students alike. The publisher hopes to create thereby both a teaching aid that gives life and depth to the study of science, as well as a significant research tool; for in those fundamental writings lie not only the seeds which have, in the meantime, developed and borne fruit, but also unnumbered other leads that await development. To those whose study and research lie in the sciences, these works offer an inexhaustible source for stimulation and for the advancement of ideas.

Today, no less than when these words were written, the proper orientation of science is toward the future. Yet, the uses of the scientific past are also becoming clearer—not the least being the continued application of Maxwell's memorable dictum on the didactic value of the study of historical accounts and of original works in science, found in the preface of the *Treatise on Electricity and Magnetism*: "It is of great advantage to the student of any subject to read the original memoirs on that subject, for science is always most completely assimilated when it is in the nascent state. . . ."

This series of connected essays on single topics of science will help us to remember that in the development of each field the overriding characteristic of scientific growth is its continuity. If science today is so good, it is to a large extent because science in the past was so good.

Gerald Holton

Jefferson Laboratory
Harvard University

Foreword

Alfred Werner was one of the truly great geniuses in the history of chemistry. The story of how the coordination theory sprang into his mind in a sudden flash of genius is well known, and the credit and honors that he received for it were well deserved. Unfortunately, many modern chemists fail to realize that many scientists preceded Werner in the study of coordination compounds and that Werner built his theory on the experimental work and theories of some of those predecessors. Anyone who has read Werner's early papers (see, for example, G. B. Kauffman, *Classics in Coordination Chemistry, Part I: The Selected Papers of Alfred Werner*, Dover Publications, Inc., New York, 1968) will realize that he was well acquainted with the chemical literature of his day; it would be foolish to believe that, in formulating his own theory, he did not consider the theories which were then under discussion in chemical circles, incomplete and faulty as they were. To concede this does not detract from the greatness of Werner's discovery, for it was he who put the pieces of the complex puzzle together and who persisted in collecting evidence to support his ideas, even in the face of severe criticism. He deserves all of the admiration and honor that he has received.

However, we often fail to recall Werner's predecessors—those who did so much to make his success possible. There were many of them, and their work stretched over nearly a century. Tassaert was the first to observe the peculiar behavior of solutions of cobalt salts when ammonia is added to them (1798). From that time on, many chemists labored in this field. They isolated and characterized "complex compounds" of cobalt, platinum, chromium, and other transition metals. Each modified the ideas of his predecessors and added new thoughts of his own. Sophus Mads Jørgensen, whose work immediately preceded Werner's and was actually contemporaneous with some of it, prepared and carefully examined many coordination compounds. Although his theory as to their structure—the so-called Blomstrand-Jørgensen chain theory—was quite erroneous, he deserves a good deal of credit, for Werner's ideas were based to a considerable extent on Jørgensen's careful laboratory work. Indeed, when Werner published his theory (1893), he had done no experimental work on complex inorganic compounds.

This book, which consists of translations of some of the papers of several of Werner's most important predecessors, should give their work the prominence that it deserves and will show that they, too, made significant contributions to the theory of complex inorganic compounds. Professor George B. Kauffman, who has collected, translated, and edited these papers, is widely recognized as the leading authority on the life and work of Alfred Werner and on the history of the coordination theory. His extensive research on these topics has been careful, painstaking, and thorough, and he has produced a volume which is both instructive and interesting. Although it will be of interest primarily to coordination chemists and historians of science, the lessons that it teaches can benefit those in every branch of chemistry. It is hoped that it will be widely read.

John C. Bailar, Jr.

University of Illinois
Urbana, Illinois

Preface

The need for a volume of *Classics in Coordination Chemistry* first became apparent to me in 1963 during my study of the life and work of Alfred Werner, the founder of coordination chemistry (*Alfred Werner, Founder of Coordination Chemistry*, Springer-Verlag, Berlin–Heidelberg–New York, 1966). The majority of the more important papers in the field were written in German and hence not readily available to most English-speaking chemists. In view of Werner's central and almost monopolistic role in coordination chemistry, I devoted an entire volume in the Classics of Science series (*Classics in Coordination Chemistry, Part I: The Selected Papers of Alfred Werner*, Dover Publications, New York, 1968) to translations of his six major papers. The present volume, which is intended to complement Part I, comprises both theoretical and experimental contributions by German, Swedish, Danish, English, and French chemists and covers the time span of 1798 to 1899, i.e., the period from the beginning of scientific coordination chemistry through the Werner–Jørgensen controversy. The six selections range in length from the 3-page excerpt from Graham's *Elements of Chemistry* to the 62-page polemic by Jørgensen.

I have followed the practice adopted in Part I of placing introductory discussions and thumbnail sketches of the authors before each individual paper rather than combining them into a single essay. My intention was to link the selections so as to approximate a review of the development of coordination chemistry—to form a continuous narrative, so to speak, or, to employ a metaphor, a motion picture rather than a series of "stills" or slides. At the same time, I hope that I have avoided giving the false but alluring impression of the progress of science as a smooth, unbroken, steadily ascending line. The history of science, in common with all history, is of course not a continuous function, but a quantized one that proceeds in forward "spurts" and occasionally leads to dead ends or backward steps. I have included here what I have considered to be the major "spurts" of coordination chemistry. Needless to say, not all of these have proved successful; the interesting case of the now defunct Blomstrand–Jørgensen chain theory, to which I have allotted much space, is an example.

Any selection of papers in a work of this sort is to some extent arbitrary, but I hope that most coordination chemists and historians of science would agree with the majority of my selections even though some might find their favorite classic missing. The failure to include such a paper was probably not an oversight, but was based upon the size limitations of this volume. Throughout the compilation of this book the need for brevity forced me time and again to write individual articles on the various topics or chemists that could not be adequately discussed here. The interested reader is therefore referred to my detailed articles in other journals and books. I regret that these digressions caused such a lengthy delay in the appearance of this volume.

In order to preserve the flavor of the original papers, I have tried consistently to retain the author's nomenclature and terminology, with addition of explanatory notes whenever necessary. My feelings about the translation process have become somewhat more liberal since I completed Part I, and the selections in the present volume should reflect this change. As a goal, I have tried to preserve "the spirit rather than the letter" of the papers, and I have not hesitated to recast unusually awkward constructions. While I readily accept full responsibility for the infelicities in the translations, perhaps it is not superfluous to point out that several of the papers were written in a language that was not the mother tongue of the author (Papers Two, Four, Five, and Six). Finally, I console myself with Alexander Pope's observation that "it is certain no literal translation can be just to an excellent original."

I wish to thank the following publishers for their kind permission to translate and/or reproduce the papers in this volume: Masson et Cie., Paris (excerpt from Tassaert's paper); Verlag Johann Ambrosius Barth, Leipzig (Papers One, Two, and Six); Lea and Febiger, Philadelphia (Paper Three); and Verlag Chemie GmbH, Weinheim/Bergstrasse (Papers Four and Five).

It is my pleasure to acknowledge the generous financial assistance of both the History and Philosophy of Science Program of the National Science Foundation Division of Social Sciences (Grant GS-74) and of the American Philosophical Society (Penrose Fund,

Grant 3255), which made possible my year's study of Werner's life and work at the Universität Zürich, where the present volume was begun.

Additional grants from NSF (Grant GS-1580) and APS (Johnson Fund, Grant 876) permitted me to continue the work during a period of trying personal vicissitudes. A sabbatical leave from the California State University, Fresno and a John Simon Guggenheim Memorial Foundation Fellowship enabled me to complete this volume a decade after its inception. Furthermore, a generous subvention grant from the Guggenheim Foundation made possible the publication of this book.

Thanks are due to Prof. Gerald Holton of Harvard University, the editor of this series, for his sound advice and unfailing support, to Prof. O. Theodor Benfey for his critical review of the manuscript, and to Prof. John C. Bailar, Jr., for writing the Foreword. I am also indebted to the California State University, Fresno Research Committee, which provided funds for typing in the form of a National Science Foundation Institutional Grant, and to Shirley Benson, Mary McIntyre, Joan Jeffries, and Elsie Taylor for typing the manuscript. I am pleased to acknowledge the assistance of the CSUF Library Reference Department, in particular of its head, William F. Heinlen, and his helpful staff members, Thomas J. Ebert, Herbert S. Fox, Diane L. Majors, and Grace T. Waibel, who often went far beyond the call of duty in locating references and biographical items. Special thanks are also due to Robert Michelotti of the CSUF Instructional Media Center for reproduction of portraits of the authors of the individual papers. I further wish to express my gratitude to my former wife, Inge, for her help in translation and for her continual encouragement. Thanks for assistance in translation are also due to Solfrid Johansen. Finally, I owe a word of apology to my wife, Laurie, and to my daughters, Ruth and Judith, for the many occasions on which I was forced to neglect them while working on this book.

G.B.K.
Visiting Scholar
Office for History of Science and Technology
University of California

Berkeley, California
July, 1976

Contents

List of Illustrations

Introduction

The field of coordination chemistry encompasses a great diversity of substances and phenomena, and in order to consider its history in a book of reasonable length, I have had to be extremely selective and to limit my discussion to what I consider the major highlights and to forgo many of the interesting but minor events. Even my treatment of these major discoveries must necessarily be brief, and rather than embark upon detailed analyses of these classics, I have preferred to let the papers speak for themselves. Of course, I have provided extensive editorial comments and annotation.

Coordination chemistry is, quite simply, the chemistry of coordination compounds. A coordination compound may be defined as a compound containing a central atom or ion to which are attached molecules or ions whose number usually exceeds the number corresponding to the oxidation number or valence of the central atom or ion. The central atom or center of coordination is usually a transition metal, that is, an element from one of the subgroups or B groups of the periodic table. The coordinated groups, which are called ligands, may be neutral molecules (such as ammonia, ethylenediamine, or water) or ions (such as chloride, bromide, oxalate, or cyanide). Metal-ammines, in which ammonia molecules are coordinated to a central metal ion, are among the commonest coordination compounds. Ligands are attached to the central atom by means of what are called coordinate bonds or coordinate covalent bonds.

As I shall emphasize throughout this volume, coordination compounds are of tremendous theoretical importance. But they are of great practical utility as well. Coordinating agents are used in metal ion sequestration or removal, solvent extraction, dyeing, leather tanning, electroplating, catalysis, water softening, and other industrial processes too numerous to mention here. In fact, new practical applications for them are found almost daily. They are of tremendous importance in biochemistry. For example, vitamin

B_{12} is a coordination compound of cobalt; the hemoglobin of our blood is a coordination compound of iron; the hemocyanin of invertebrate animal blood is a coordination compound of copper; and the chlorophyll of green plants is a coordination compound of magnesium. Historically, the accumulation of information and experimental data about these compounds has been a very slow and gradual process.

Some Historically Significant Coordination Compounds

Different historians ascribe different dates to the discovery of the first coordination compound. If we consider alizarin dyes as coordination compounds, we can probably date them from biblical times with Joseph's "coat of many colors."[1] The bright red alizarin dye, a calcium aluminum chelate compound of hydroxyanthraquinone, was first used in India and was known to the ancient Persians and Egyptians long before it was used by the Greeks and Romans.[2] Probably the first scientifically recorded observation of a completely inorganic coordination compound is Andreas Libavius' description[3] in 1597 of the formation in solution of the familiar blue tetraamminecopper(II) ion, $[Cu(NH_3)_4]^{2+}$. Another candidate for the first coordination compound is Prussian blue, a complex of empirical formula $KCN \cdot Fe(CN)_2 \cdot Fe(CN)_3$ first obtained accidentally in 1704 by Diesbach, a manufacturer of artist's colors from Berlin.[4]

If we limit ourselves to the cobalt(III)-ammines, the most familiar type of coordination compound, the question of priority receives a more definitive answer. Most authorities attribute the discovery of the first metal-ammine to an otherwise unknown Citizen Tassaert (no first name given).[5] Some chemists imply or

---

[1] Genesis, 37:3.

[2] D. F. Martin and B. B. Martin, *Coordination Compounds*, McGraw-Hill Book Co., Inc., New York, 1964, Chap. 2.

[3] A. Libavius, *Commentationvm Metallicarvm Libri Qvatvor de Natvra Metallorvm . . .*, Frankfurt, 1597, vi, Part II, Chap. 19.

[4] Anonymous, *Miscellanea Berolinensia*, *1*, 377 (1710).

[5] In the *Royal Society Catalogue of Scientific Papers*, *5*, 914 (1871), Tassaert's initials are given as B.M., but his initials are not given in any of his three papers in *Annales de Chimie.* This journal was controlled by a "company," of which Tassaert was a member. Antoine-François Fourcroy (*Système des connaissances chimiques*, Paris, 1801, Vol. 5, pp. 108–109) identifies Tassaert as "chimiste

even openly state that Tassaert was the first to prepare and isolate hexaamminecobalt(III) chloride, the parent compound from which all cobalt-ammines may be considered to be derived. Yet, in 1798 he merely observed the brownish mahogany color of the hexaamminecobalt(III) ion, $[Co(NH_3)_6]^{3+}$, formed when excess aqueous ammonia is added to a solution of cobalt chloride or nitrate, and he failed to follow up his accidental discovery.[6]

Tassaert's article is entitled *Analyse du Cobalt de Tunaberg, suivie de plusieurs moyens d'obtenir ce métal à l'état de pureté, et de quelques-unes de ses propriétés les plus remarquables* and deals with the dissolution and analysis of a cobalt ore from Tunaberg (Kolmården, Södermanland, Sweden). Tassaert used excess ammonia to dissolve the cobalt and simultaneously to precipitate the iron in the ore. The passage of interest to coordination chemists occurs on pp. 106–107:

> Another rather surprising phenomenon is that when nitrate of cobalt is precipitated by excess ammonia, there is formed a precipitate[7] which immediately redissolves and gives a brown solution; but if this solution is diluted at once with a large amount of water, there is formed a green precipitate which consists only of pure oxide of cobalt, which dissolves in acids and imparts a beautiful pink color to the solutions; if on the contrary this solution of cobalt in ammonia is left exposed to the air for a long time, it can be diluted with as much water as one wishes without the formation of any precipitate. This fact, which I do not yet claim to explain, but which I intend to reconsider, has appeared to me to be worthy of recording here.

Tassaert's chance observation was soon followed by additional discoveries. Among historically significant coordination compounds discovered in the first half of the nineteenth century we may list tetraamminepalladium(II) tetrachloropalladate(II) (Vauquelin's Salt), $[Pd(NH_3)_4][PdCl_4]$ (1813);[8] hexaamminecobalt(III) oxalate, $[Co(NH_3)_6]_2(C_2O_4)_3$ (1822);[9] potassium ferricyanide, $K_3[Fe(CN)_6]$ (1822);[10] potassium hexacyanocobaltate(III), $K_3[Co(CN)_6]$

prussien chargé des travaux du laboratoire de l'école des mines à Paris." I am indebted for this meager information to Profs. William A. Smeaton and Maurice P. Crosland and to Mrs. Susan Court. Authorities in Paris were unable to provide any information.

[6] Citoyen Tassaert, *Ann. chim. phys.*, *28*, 92–112 (1798).

[7] On p. 95 of the article Tassaert describes the precipitate as rose colored.

[8] L.-N. Vauquelin, *Ann. chim. phys.*, *88*, 188 (1813).

[9] L. Gmelin, *Handbuch der theoretischen Chemie*, 2nd ed., Varrentrapp, Frankfurt am Main, 1822, Vol. 2, p. 928.

[10] L. Gmelin, *Schweigger's J. Chem. Physik*, *34*, 325 (1822).

(1822);[11] potassium tetracyanoplatinate(II), $K_2[Pt(CN)_4]$ (1822);[12] potassium trichloro(ethylene)platinate(II) monohydrate (Zeise's Salt), $K[Pt(C_2H_4)Cl_3]\cdot H_2O$ (1827) (Paper Two of this volume);[13] tetraammineplatinum(II) tetrachloroplatinate(II) (Magnus' Green Salt), $[Pt(NH_3)_4][PtCl_4]$ (1828)[14] (Paper One); *cis*-dichlorodiammineplatinum(II) (Peyrone's Salt), *cis*-$[Pt(NH_3)_2Cl_2]$ (1844);[15] *trans*-dichlorodiammineplatinum(II) (Reiset's Second Chloride), *trans*-$[Pt(NH_3)_2Cl_2]$ (1844);[16] aquopentaamminecobalt(III) salts (roseo salts), $[Co(NH_3)_5H_2O]X_3$ (discovered in 1847, but not published until 1851);[17] and chloropentaamminecobalt(III) salts (purpureocobalt salts), $[Co(NH_3)_5Cl]X_2$ (1851).[18]

Most of these discoveries, however, were isolated instances. In July of 1852 Oliver Wolcott Gibbs (1822–1908),[19] an American chemist, began a collaboration with Frederick Augustus Genth (1820–1893),[20] a German chemist and mineralogist who had emigrated to the United States. The results of their painstaking research did not appear until 1856 in a 67-page monograph entitled *Researches on the Ammonia-Cobalt Bases*,[21] which has since become famous in the annals of coordination chemistry. In this elaborate and extensive memoir, which "has always ranked among the highest chemical investigations ever made in this country",[20a] Gibbs and Genth described in detail the preparation, properties, analytical

---

[11] L. Gmelin, *Schweigger's J. Chem. Physik, 36,* 230 (1822).

[12] L. Gmelin, *Handbuch der theoretischen Chemie*, 2nd ed., Varrentrapp, Frankfurt am Main, 1822, Vol. 2, p. 1692.

[13] W. C. Zeise, *Ann. Phys. Chem., 9,* 632 (1827); *ibid., 21,* 497–541 (1831).

[14] G. Magnus, *Ann. Phys. Chem., 14,* 239–242 (1828).

[15] M. Peyrone, *Ann., 51,* 1–29 (1844).

[16] J. Reiset, *Compt. rend., 18,* 1100 (1844); *Ann. chim. phys.,* [3] *11,* 417–433 (1844).

[17] F. A. Genth, *Keller-Tiedemann's Nordamerikanischer Monatsbericht für Natur- und Heilkunde, 2,* 8–12 (1851).

[18] F. Claudet, *Phil. Mag., 2,* 253–260 (1851).

[19] C. S. Pierce, *The Evening Post*, New York, Dec. 10, 1908; T. W. Richards, *Ber., 42,* 5037–5054 (1909); *idem, Science, 29,* 101–103 (1909); F. W. Clarke, *J. Chem. Soc., 95,* 1299–1312 (1909); *idem, Biog. Mem., Nat. Acad. Sci., 7,* 1–22 (1910); C. L. Jackson, *Am. J. Sci., 28,* 253–259 (1909); C. E. Munroe, *Science, 34,* 864–868 (1911); G. B. Kauffman, *Platinum Metals Review, 16* (3), 101–104 (1972).

[20] G. F. Barker, (*a*) *Proc. Amer. Phil. Soc., 40,* x–xxii (1901); *idem, Biog. Mem., Nat. Acad. Sci., 4,* 201–231 (1901); (*b*) W. M. Meyers and S. Zerfoss, *J. Franklin Inst., 241,* 341–354 (1946); (*c*) G. B. Kauffman, in *Dictionary of Scientific Biography*, ed. by C. C. Gillispie, Charles Scribner's Sons, New York, 1972, Vol. 5, pp. 349–350.

[21] O. W. Gibbs and F. A. Genth, *Researches on the Ammonia-Cobalt Bases,* Smithsonian Institution, Washington, 1856; reprinted in *Am. J. Sci.,* [2] *23,* 235–265, 319–341 (1856); *ibid.,* [2] *24,* 86–101 (1857).

data, and reactions of 35 salts of four cobalt-ammine cations (aquopentaamminecobalt(III) (roseocobalt), $[Co(NH_3)_5H_2O]^{3+}$; chloropentaamminecobalt(III) (purpureocobalt), $[Co(NH_3)_5Cl]^{2+}$; hexaamminecobalt(III) (luteocobalt), $[Co(NH_3)_6]^{3+}$; and nitropentaamminecobalt(III) (xanthocobalt), $[Co(NH_3)_5NO_2]^{2+}$) and simultaneously set the pattern for the systematic comprehensive studies of coordination compounds later undertaken by investigators such as Sophus Mads Jørgensen and Alfred Werner.[22]

Theories of Coordination Compounds

In most fields of science, theory lags behind practice. Sufficient experimental data must be accumulated before attempts are made to explain these experimental facts and to predict new phenomena. As we have just seen, during the first half of the nineteenth century discoveries of coordination compounds were few, sporadic, and often accidental, and it was not until after Gibbs and Genth's classic memoir of 1856 that chemists began to devote themselves in earnest to a systematic study of this field. We might therefore think that few theories of coordination compounds were advanced until late in the second half of the nineteenth century, but this was not the case. In coordination chemistry, the lag of theory behind practice was not a great one because of the tremendous importance of coordination compounds to the general question of chemical bonding. In the words of Alfred Werner himself, "Of all inorganic compounds, [metal-ammines] are best suited to the solution of constitutional problems. . . . it was through the investigation of metal-ammines that the decisive basic principles involved in the constitutional conception of inorganic compounds could first of all be clearly recognized."[23]

During the nineteenth century theories of coordination compounds[24] were advanced by Thomas Graham (1837) (Paper Three),[25] Jöns Jacob Berzelius (1841),[26] Jules Reiset (1844),[16]

---

[22] G. B. Kauffman, *Isis*, *67* (1977), in press.

[23] A. Werner, *Ber.*, *40*, 15–69 (1907).

[24] For discussions of some of these theories see F. Reitzenstein, *Z. anorg. Chem.*, *18*, 152–210 (1898), and J. C. Bailar, Jr., "The Early Development of the Coordination Theory," in *The Chemistry of the Coordination Compounds*, J. C. Bailar, Jr., ed., Reinhold Publishing Corp., New York, 1956, Chap. 2.

[25] T. Graham, *Elements of Chemistry*, London, 1837; *Lehrbuch der Chemie*, transl. by F. J. Otto, F. Vieweg und Sohn, Braunschweig, 1840, Vol. 2, p. 741.

[26] J. J. Berzelius, *Jahresbericht*, *21*, 108 (1841).

Charles Frédéric Gerhardt (1850),[27] August Wilhelm von Hofmann (1851),[28] Carl Ernst Claus (1856) (Paper Four),[29] Friedrich August Kekulé (1864),[30] Christian Wilhelm Blomstrand (1869–1871) (Paper Five)[31] (subsequently modified and developed by Sophus Mads Jørgensen, Paper Six),[32] and finally, Alfred Werner (1893),[33] whose coordination theory is today the universally accepted explanation for the constitution, properties, and reactions of complex compounds. These theories will be discussed in more or less detail later in this volume.

---

[27] C. Gerhardt, *Jahresbericht*, *3*, 335 (1850).

[28] A. W. Hofmann, *Ann.*, *78*, 253 (1851).

[29] C. Claus, *Ann.*, *98*, 317–333 (1856).

[30] A. Kekulé, *Compt. rend.*, *58*, 510 (1864).

[31] C. W. Blomstrand, *Die Chemie der Jetztzeit*, Carl Winter's Universitätsbuchhandlung, Heidelberg, 1869, pp. 280 ff.; *Ber.*, *2*, 202–206 (1869); *ibid.*, *4*, 40–52 (1871).

[32] S. M. Jørgensen, *Z. anorg. Chem.*, *19*, 109–157 (1899).

[33] A. Werner, *Z. anorg. Chem.*, *3*, 267–330 (1893). For a discussion and English translation of this paper and five others by Werner see G. B. Kauffman, *Classics in Coordination Chemistry, Part I: The Selected Papers of Alfred Werner*, Dover Publications, Inc., New York, 1968.

INTRODUCTION TO PAPER ONE[1]

[When naming coordination compounds, early chemists were like the children of Israel, who called the food miraculously supplied to them in the wilderness *manna*, "for they wist not what it was" (Exodus 16:15). Since the true constitutions and configurations of these compounds were unknown until the advent of Alfred Werner's coordination theory,[2,3] complexes were often named after their colors (luteo, purpureo, roseo, praseo, violeo, etc.)[4] or after the discoverers (Cleve's Salt, Erdmann's Salt, Fischer's Salt, Recoura's Sulfate, Reinecke's Salt, etc.). In 1897 Werner[5] proposed a systematic nomenclature for cobalt-ammines based on his coordination theory, and this system is used today with only minor modifications.

One of the most famous and important coordination compounds named after its discoverer is Magnus' Green Salt, tetraammineplatinum(II) tetrachloroplatinate(II), $[Pt(NH_3)_4][PtCl_4]$, discovered in 1828 by Gustav Magnus (Paper One).[6] As the first discovered platinum-ammine, it is numbered among the earliest known coordination compounds. It is true that the corresponding pink palladium compound, tetraamminepalladium(II) tetrachloropalladate(II), $[Pd(NH_3)_4][PdCl_4]$ (Vauquelin's Salt), was discovered fifteen years earlier,[7] but Magnus' Salt has played a much more significant role in the development of coordination chemistry because of the great stability and retention of configuration characteristic of platinum compounds. Magnus' Salt later served as the starting point for a great amount of research on the so-called platinum bases carried out by chemists such as Gros, Reiset, Peyrone, Raewsky, Gerhardt, Cleve, and many others. It is still a valuable intermediate and useful reagent for the preparation of a wide variety of platinum complexes.

Magnus added excess aqueous ammonia to a solution of platinum(II) chloride in hydrochloric acid, whereupon a green, insoluble salt was

---

[1] Material in square brackets is supplied by the editor of this volume.

[2] A. Werner, *Z. anorg. Chem.*, *3*, 267–330 (1893). For a discussion and English translation of this paper see G. B. Kauffman, *Classics in Coordination Chemistry, Part I: The Selected Papers of Alfred Werner*, Dover Publications, Inc., New York, 1968, pp. 5–88.

[3] G. B. Kauffman, *Alfred Werner, Founder of Coordination Chemistry*, Springer-Verlag, Berlin–Heidelberg–New York, 1966; *Chemistry*, *39* (12), 14–18 (1966); *Educ. Chem.*, *4*, 11–18 (1967).

[4] E. Fremy, *Ann. chim. phys.*, [3] *35*, 257 (1852); *J. prakt. Chem.*, *57*, 95 (1852).

[5] A. Werner, *Z. anorg. Chem.*, *14*, 21–27 (1897).

[6] G. Magnus, *Ann. Phys. Chem.*, *14*, 239–242 (1828).

[7] L.-N. Vauquelin, *Ann. chim. phys.*, *88*, 188 (1813); G. B. Kauffman and J. H. Tsai, *Inorg. Syn.*, *8*, 234–238 (1966).

slowly deposited. With the formation of this precipitate, Magnus simultaneously discovered both the most important cation ($[Pt(NH_3)_4]^{2+}$) and the most important anion ($[PtCl_4]^{2-}$) of divalent platinum. More convenient, alternative syntheses have since been devised.[8] Magnus also described the preparation of the tetrachloroplatinates of potassium, ammonium, and sodium,[9] and most of his short but historic article is devoted to these compounds.

Magnus' Green Salt has the same empirical formula as several other platinum(II)-ammine compounds, viz., *cis*-$[Pt(NH_3)_2Cl_2]$ (Peyrone's Chloride),[8b,9b] *trans*-$[Pt(NH_3)_2Cl_2]$ (Reiset's Second Chloride),[9b,10] $[Pt(NH_3)_3Cl][Pt(NH_3)Cl_3]$,[11] $[Pt(NH_3)_4][Pt(NH_3)Cl_3]_2$,[12] and $[Pt(NH_3)_3Cl]_2[PtCl_4]$.[11,13]

The series constitutes a classic example of what Werner has called "coordination polymerism" (*Koordinationspolymerie*),[14,15] a type of structural isomerism in which the isomers have the same composition but formula weights that are different multiples of the same formula weight. The last compound in the series is believed to be one of the constituents of a so-called red modification of Magnus' Green Salt with almost identical physical and chemical properties, discovered by Jørgensen and Sørensen[16] and subsequently investigated by others.[17] Peyrone[8b] was the first to formulate Magnus' Green Salt correctly as $[Pt(NH_3)_4][PtCl_4]$.

The fact that Magnus' Green Salt differs markedly in color from its constituent ions ($[Pt(NH_3)_4]^{2+}$, colorless; $[PtCl_4]^{2-}$, orange-red) indicates considerable interaction between the cationic metal atom and anionic metal atom, and consequently its crystal structure and related

---

[8] (*a*) J. Gros, *Ann.*, *27*, 241 (1838); *Ann. chim. phys.*, [2] *69*, 204 (1838); (*b*) M. Peyrone, *Ann.*, *51*, 1 (1844); (*c*) C. Claus, *ibid.*, *107*, 138 (1858); (*d*) F. W. Clarke and M. E. Owens, *Am. Chem. J.*, *3*, 350 (1881–82); (*e*) I. I. Chernyaev, *Izvest. Inst. Izucheniiu Platiny i Drugikh Blagorodn. Metal.*, *7*, 52 (1929); (*f*) R. N. Keller, *Inorg. Syn.*, *2*, 250–253 (1946).

[9] (*a*) R. N. Keller, *Inorg. Syn.*, *2*, 247–250 (1946); (*b*) G. B. Kauffman and D. O. Cowan, *ibid.*, *7*, 239–245 (1963).

[10] J. Reiset, *Ann. chim. phys.*, [3] *11*, 417–433 (1844); *Compt. rend.*, *18*, 1100–1105 (1844).

[11] M. Peyrone, *Ann.*, *55*, 205 (1845).

[12] A. Cossa, *Gazz. chim. ital.*, *20*, 725 (1890); *Ber.*, *23*, 2503–2509 (1890).

[13] P. T. Cleve, *Svenska Akad. Handl.*, *10*, 64 (1872); L. A. Chugaev, *J. Chem. Soc.*, *107*, 1247 (1915); H. J. S. King, *ibid.*, *1948*, 1912–1916.

[14] A. Werner, *Neuere Anschauungen auf dem Gebiete der anorganischen Chemie*, Friedrich Vieweg, Braunschweig, 1st ed., 1905, pp. 159–162; 2nd ed., 1909, pp. 252–256; 3rd ed., 1913, pp. 319–324; 4th ed., 1920, pp. 328–333; 5th ed., 1923, pp. 330–335.

[15] G. B. Kauffman, *Coord. Chem. Revs.*, *11*, 161–188 (1973).

[16] S. M. Jørgensen and S. P. L. Sørensen, *Z. anorg. Chem.*, *48*, 441–445 (1906).

[17] (*a*) E. G. Cox, F. W. Pinkard, W. Wardlaw, and G. H. Preston, *J. Chem. Soc.*, *1932*, 2527–2535; (*b*) H. D. K. Drew and H. J. Tress, *ibid.*, *1935*, 1586–1588.

properties have been extensively investigated.[18] The salt is perhaps the simplest prototype of a large class of d^8 metal complexes exhibiting short metal-metal spacings.[18c] The absorption spectra[18b] and electronic spectra[19] differ greatly from those of its constituent ions. The crystal structures of its cation[20] and anion[21] have both been determined, and in each case the central platinum atom has been found to be surrounded by four ammonia molecules of four chlorine atoms, respectively, in the square planar arrangement predicted by Werner:[2]

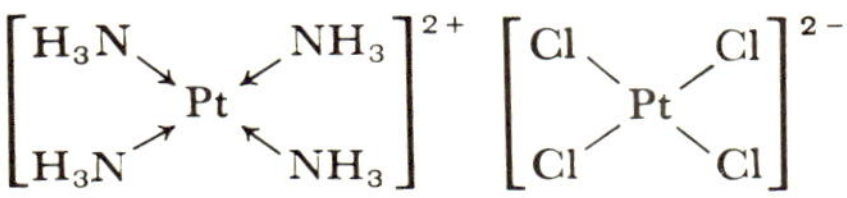

In the crystal these cations and anions are stacked with parallel planes forming continuous chains of metal atoms.[17a,20e,22]

In 1966 Collman[23] suggested that single crystals of transition metal complexes with linear chains of metal atoms might behave as anisotropic metallic conductors. Single crystals of Magnus' Green Salt have been found to exhibit anisotropic semiconductor properties and photoconduction with the highest conduction in the direction of the metal-metal chain, indications of considerable direct metal-metal interaction.[24] The unusual solid state properties associated with this unusual compound will un-

---

[18] (*a*) E. Hertel and K. Schneider, *Z. anorg. Chem.*, *202*, 77–80 (1931); (*b*) S. Yamada, *J. Am. Chem. Soc.*, *73*, 1579–1580 (1951); (*c*) B. A. Scott and F. Mehran, *165th National Meeting, American Chemical Society, Dallas, Texas, April, 1973,* Inorganic Abstract 129.

[19] J. R. Miller, *J. Chem. Soc.*, *1965*, 713–720; P. Day, A. F. Orchard, A. J. Thompson, and R. J. P. Williams, *J. Chem. Phys.*, *42*, 1973–1981 (1965); *43*, 3763–3764 (1965); B. G. Anex, M. E. Ross, and M. W. Hedgcock, *ibid.*, *46*, 1090–1095 (1967).

[20] (*a*) H. D. K. Drew, F. W. Pinkard, W. Wardlaw, and E. G. Cox, *J. Chem. Soc.*, *1932*, 1004–1015; (*b*) E. G. Cox, *ibid.*, *1932*, 1912–1920; (*c*) E. G. Cox and G. H. Preston, *ibid.*, *1933*, 1089–1093; (*d*) B. N. Dickinson, *Z. Krist.*, *88*, 281–297 (1934); (*e*) N. V. Belov, *Izvest. Inst. Izucheniiu Platiny i Drugikh Blagorodn. Metal.*, *18*, 112 (1945).

[21] R. G. Dickinson, *J. Am. Chem. Soc.*, *44*, 2404–2411 (1922); L. Pauling and M. L. Huggins, *Z. Krist.*, *87*, 205–238 (1934), pp. 205, 232.

[22] J. P. Collman, L. F. Ballard, L. K. Monteith, C. G. Pitt, and L. Slifkin, in *International Symposium on Decomposition of Organometallic Compounds to Refractory Ceramics, Metals, and Metal Alloys*, ed. by K. S. Mazdiyasni, University of Dayton Research Institute, Dayton, Ohio, 1968, pp. 269–283.

[23] C. G. Pitt, L. K. Monteith, L. F. Ballard, J. P. Collman, J. C. Morrow, W. R. Roper, and D. Ulkü, *J. Am. Chem. Soc.*, *88*, 4286–4287 (1966); J. P. Collman, *Chem. Eng. News*, *45*, 50–52 (Dec. 11, 1967).

[24] L. V. Interrante and F. P. Bundy, *Inorg. Chem.*, *10*, 1169–1174 (1971); L. V. Interrante and R. P. Messmer, *ibid.*, *10*, 1174–1180 (1971); P. S. Gomm, T. W. Thomas, and A. E. Underhill, *J. Chem. Soc.*, *A*, *1971*, 2154–2157.

doubtedly make it an object of even more experimental and theoretical studies in the future.

HEINRICH GUSTAV MAGNUS,[25] Professor of Physics and Technology at the University of Berlin, is as well known for his contributions to these fields as for those to chemistry. Born in Berlin on May 2, 1802, the son of Johann Matthias Magnus, the prosperous founder of a large trading firm, Magnus received private instruction in mathematics and natural science. He entered the University of Berlin in 1822, and in 1825 he published his first paper, an investigation of pyrophoric iron, cobalt, and nickel[26] carried out under the direction of Eilhard Mitscherlich (1794–1863), discoverer of the law of isomorphism. Following his attainment of the doctorate in September, 1827, with a dissertation on tellurium, Magnus took the advice of Mitscherlich, Heinrich Rose (1795–1864), Gustav Rose (1798–1873), and Friedrich Wöhler (1800–1882), who had all studied with Jöns Jacob Berzelius (1779–1848), and in October, 1827, went to Stockholm to study with the Master, who became his lifelong friend and adviser.[27]

It was in Berzelius' laboratory that Magnus not only discovered the first platinum-ammine compound (Magnus' Green Salt, $[Pt(NH_3)_4]$-$[PtCl_4]$) and its related potassium salt ($K_2[PtCl_4]$) but also worked on the addition compound of ethylene and platinous chloride later described by the Danish chemist William Christoffer Zeise (1789–1847) (Zeise's Salt, $K[Pt(C_2H_4)Cl_3] \cdot H_2O$) (Paper Two). In the summer of 1828 Magnus returned to Berlin, where, with the exception of a visit to Gay-Lussac's laboratory in Paris during 1828 and 1829, he remained until his death on April 4, 1870.

Magnus' *Habilitationsschrift* on mineral analysis (1831) permitted him to begin lecturing on technology at the university and on chemistry at the Städtische Gewerbeschule (Municipal Technical School) but caused a break with his teacher Mitscherlich, who regarded the young *Privat-Dozent* as a dangerous competitor. In 1833 Magnus was appointed *Ausserordentlicher* (*Extraordinarius*) *Professor* and in 1845 *Ordentlicher* (*Ordinarius*) *Professor* of Physics and Technology at the University of Berlin, where he also served as *Rektor* during 1861 and 1862. Magnus married Bertha Humbolt in 1840. He became a member of the Berliner Akademie

[25] A. W. Hofmann, *Ber.*, *3*, 993–1101 (1870); A. W. Williamson, *J. Chem. Soc.*, *24*, 610–615 (1871); W. Prandtl, *Deutsche Chemiker in der ersten Hälfte des neunzehnten Jahrhunderts*, Verlag Chemie, Weinheim/Bergstrasse, 1956, pp. 303–314; G. B. Kauffman, *Dictionary of Scientific Biography*, ed. by C. C. Gillispie, Charles Scribner's Sons, New York, 1974, Vol. 9, pp. 18–19; G. B. Kauffman, *Platinum Metals Review*, *20*(1), 21–24 (*1976*).

[26] G. Magnus, *Ann. Phys. Chem.*, *79*, 81 (1825).

[27] E. Hjelt, ed., *Aus Jac. Berzelius und Gustav Magnus' Briefwechsel in den Jahren 1828–1847*, F. Vieweg und Sohn, Braunschweig, 1900.

der Wissenschaften in 1840 and was one of the organizing members of the Deutsche Chemische Gesellschaft, founded in 1868. A number of his students (Wilhelm Beetz, Hermann Helmholtz, Gustav Wiedemann, Robert Schneider, John Tyndall, Rudolph Weber, and F. H. Adolph Wüllner) became famous physicists in their own right.

As was true of most chemists of the time, Magnus' research interests were varied. From an initial interest in mineral analysis, he turned to inorganic chemistry (discovered periodic acid and its salts in 1833),[28] organic chemistry (discovered ethionic acid, $HSO_3 \cdot CH_2 \cdot CH_2 \cdot SO_3H$, and isethionic acid, $CH_2OH \cdot CH_2 \cdot SO_3H$, in 1833–1839;[29] polymerization of hydrocarbons on heating, 1853), physiological chemistry (oxygen and carbon dioxide content of blood, 1837–1845), and agricultural chemistry (1849). He gradually turned more and more to physicochemical and eventually purely physical investigations, which constitute his most important scientific achievements. Among these we may cite his contributions to the theory of heat, thermal expansion of gases, transmission of heat through gases by conduction and radiation, diathermaneity of gases, boiling of liquids, vapor formation, electrolysis ("Magnus' Law," which states that for each metal there is a specific voltage at which it is deposited from a solution containing a mixture of metallic salts),[30] induced and thermoelectric currents, optics, hydrodynamics, magnetism, and mechanics. Although Magnus' main claim to fame was as a physicist, he never ceased to work on chemical problems in his private laboratory. These later chemical works, however, never matured to results of general significance; they served merely for his own instruction.

Gustav Magnus was neither a theoretician nor an original thinker. The objects of his work had almost all been investigated by other scientists. He was, however, an acute, conscientious, and diligent experimenter, who uncovered much valuable physical and chemical data. He is best known to chemists for his discovery of the important green salt that bears his name.—G.B.K.]

---

[28] G. Magnus and C. F. Ammermüller, *Ann. Phys. Chem.*, *104*, 514 (1833); *Ann. chim.*, *53*, 92 (1833).
[29] G. Magnus, *Ann. Phys. Chem.*, *27*, 367 (1833); *ibid.*, *47*, 509 (1839); *Ann. Pharm.*, *6*, 152 (1833); *ibid.*, *32*, 249 (1839).
[30] G. Magnus, *Ann. Phys. Chem.*, *102*, 1 (1857).

1

Gustav Magnus

Some Compounds of Platinous Chloride

[Translation of "Ueber einige Verbindungen des Platinchlorürs," Pogendorff's *Annalen der Physik und Chemie*, *14*, 239–242 (1828).]

Until now, two compounds of platinum with chlorine have been known. One, with four atoms of chlorine, platinic chloride [*Platinchlorid*] $PtCl^4$, is obtained when platinum is dissolved in aqua regia. The other, with two atoms of chlorine, platinous chloride [*Platinchlorür*] $PtCl^2$, is obtained, on the other hand, when platinic chloride is heated to the melting point of lead, whereupon two atoms of chlorine escape.

As is known, platinic chloride readily forms double salts[1] with potassium chloride, sodium chloride, ammonium chloride, and as Bonsdorff[2,3] reports, with other metal chlorides also. Of these, the potassium salt, KCl^2+PtCl^4,[4] has been recently reinvestigated by Berzelius[5,6] and has been used for the determination of the atomic weight of platinum, which by this means was found to be 1233.26.[7]

---

[1] [In modern terminology, hexachloroplatinates(IV). The anions in these salts are true complexes, not to be confused with double salts that decompose in solution into their constituent ions, e.g., the alums.—G.B.K.]

[2] [Pehr Adolf von Bonsdorff (Abo [Turku], Finland, Oct. 27, 1791—Helsingfors [Helsinki], Finland, Jan. 11, 1839) was assistant (1816) and then professor (1823) at Helsingfors.—G.B.K.]

[3] *Ann. chim. phys.*, *34*, 150 [(1800)]; *Ann.* [*Phys.*], *11*, 124 [(1828)].

[4] [This "pre-Karlsruhe" formula corresponds to $K_2[PtCl_6]$, potassium hexachloroplatinate(IV) (potassium chloroplatinate).—G.B.K.]

[5] [Jöns Jacob Berzelius (Väversunda, Sweden, Aug. 20, 1779—Stockholm, Sweden, August 7, 1848), in whose laboratory Magnus worked during 1827 and 1828, was one of the most prominent and influential chemists of the first half of the nineteenth century.—G.B.K.]

[6] *Ann.* [*Phys.*], *13*, 469 [(1828)].

[7] [197.2 on the scale of O = 16; Magnus used O = 100. The atomic weight of platinum is now taken as 195.09.—G.B.K.]

On the other hand, platinous chloride has until now been less well known, although it forms double compounds similar to those of platinic chloride. I wish to describe briefly a few of these here.

Pure platinous chloride is insoluble in water; however, if the platinic chloride from which it has been prepared has not been heated sufficiently,[8] the residue dissolves in water with so dark brown a color that the clear liquid appears completely opaque. This liquid is a solution of platinous chloride in platinic chloride. If this is evaporated, there is precipitated a brown powder, which, when the liquid has been decanted, is no longer soluble in water. However, if the decanted liquid is again added to the water, the brown precipitate dissolves on heating.

However, this differs from the platinous chloride obtained by heating platinic chloride, not only in its color but especially in the fact that it dissolves much more easily in hydrochloric acid than does that platinous chloride, but aside from this, both have the same composition. After the brown precipitate had been washed with water until the precipitate was pure, it was reduced with hydrogen gas; it lost 25.95 per cent chlorine. On the other hand, the platinous chloride which had been washed several times with hydrochloric acid lost on similar treatment 26.38 per cent. According to calculations, platinous chloride should contain 26.41 per cent chlorine.[9] The difference between the two results is certainly small enough to be ascribed to the unavoidable errors in observation.

It appears, however, that the amount of platinous chloride dissolved by the platinic chloride depends upon the concentration of the platinic chloride solution. For if the brown solution of platinous chloride in platinic chloride is evaporated, then, as has already been noted, platinous chloride is precipitated; if it is evaporated to dryness, then although by continuous boiling the entire mass can of course be redissolved in water, the platinous chloride remains undissolved if only cold water is used, and a compound of platinic chloride with less platinous chloride dissolves. By repeated treatment of this kind the solution can be freed more and more of

---

[8] [At that time the usual method for preparing $PtCl_2$ involved heating powdered $H_2[PtCl_6]\cdot 6H_2O$, usually obtained by dissolving platinum in *aqua regia* and designated as $PtCl^4$, and freeing the resultant product from undecomposed starting material by treatment with hot water (J. J. Berzelius, *Ann. chim.*, *83*, 168 (1812); *Svenska Akad. Handl.*, *1813*, 196; *J. Chem. und Phys.*, *7*, 56 (1813), and E. Davy, *Phil. Mag.*, *40*, 272 (1812)).—G.B.K.]

[9] [26.66%, using the latest atomic weights.—G.B.K.]

platinous chloride, and therefore I do not believe that we are dealing here with a compound of two degrees of chlorination but only with a solution of one in the other.

The dissolving of this brown lower chloride [*Chlorür*] in hydrochloric acid appears to me to be quite different. This process is accompanied by formation of a peculiar red color regardless of the quantity, so that I would like to believe that there is formed here a chemical compound of platinous chloride with hydrogen chloride[10] exactly like those which Boullay[11,12] assumes are formed by other metal chlorides and iodides with hydrogen chloride and hydrogen iodide.

If potassium chloride is added to such a solution of platinous chloride in hydrochloric acid, prisms of a red salt, which is a compound of platinous chloride with potassium chloride, crystallize out containing no water of crystallization.

I have analyzed this salt by reducing a weighed amount of it with hydrogen.[13] The loss in weight represented the chlorine which had been combined with the platinum. The residue consisted of metallic platinum and potassium chloride; the latter was extracted with water, then the platinum was weighed, and the potassium chloride determined from the loss in weight. In this way I found the salt to be composed of:

Platinum	46.74	[%]
Chlorine	17.47	[%]
Potassium chloride	35.79	[%]
	100.00	[%]

If the salt is assumed to be $PtCl^2 + KCl^2$[14] and this composition is calculated accordingly, one obtains:

Platinum	47.28	[%]
Chlorine	16.97	[%]
Potassium chloride	35.75	[%]
	100.00[15]	[%]

[10] [$H_2[PtCl_4]$, tetrachloroplatinic(II) acid or chloroplatinous acid.—G.B.K.]

[11] [Pierre François Guillaume Boullay (Caen, France, April 27, 1777—Paris, France, Nov. 3, 1869).—G.B.K.]

[12] *Ann. chim. phys.*, *34*, 337 [(1800)]; *Ann.* [*Phys.*], *11*, 99 [(1828)].

[13] [$K_2[PtCl_4] + H_2 \rightarrow Pt + 2KCl + 2HCl\uparrow$.—G.B.K.]

[14] [This "pre-Karlsruhe" formula corresponds to $K_2[PtCl_4]$, potassium tetrachloroplatinate(II) or potassium chloroplatinite.—G.B.K.]

[15] [According to the latest atomic weights, the composition should be Pt 47.01%, Cl 17.08%, and KCl 35.91%.—G.B.K.]

This salt is readily soluble in water but almost insoluble in alcohol; therefore if alcohol is added to an aqueous solution of the salt, it is precipitated as delicate, rose-red, crystalline fibers.[16]

The double salts of platinous chloride with sodium chloride and ammonium chloride are obtained in a manner similar to the potassium salt. Since the compound with sodium chloride is very readily soluble in alcohol, I could not obtain it in crystalline form; on the other hand, the ammonium salt crystallizes very easily in dark red crystals.[17]

If caustic ammonia is added in excess to a solution of platinous chloride in hydrochloric acid, a green salt is precipitated after some time, which is completely crystalline and which contains platinum, chlorine, and ammonia.[18] In order to analyze it, a weighed amount of it was heated in a small retort with pure sodium carbonate [*kohlensaures Natron*];[19] the escaping gases were conducted into a solution of silver nitrate [*salpetersaures Silber*] in order to be certain that no hydrochloric acid was carried off. The resulting sodium chloride and excess sodium carbonate were extracted from the residue with water, and the chlorine was precipitated and determined with silver [nitrate] solution; the residual metallic platinum was also weighed. In addition, the quantity of the latter was determined by reduction with hydrogen gas. The amount of ammonia was calculated by difference. In this way I found this salt to be composed of:

Platinum	65.65	[%]
Chlorine	23.46	[%]
Ammonia	10.89	[%]

If the same lower chloride of platinum with ammonia is assumed

[16] [$K_2[PtCl_4]$ is now usually prepared by the reduction of $K_2[PtCl_6]$ in aqueous suspension or by the reduction of $H_2[PtCl_6]$ to $H_2[PtCl_4]$ followed by precipitation with KCl solution. Various reducing agents, such as $K_2C_2O_4$ (R. N. Keller, *Inorg. Syn.*, *2*, 247–250 (1946)) or $N_2H_4 \cdot 2HCl$ (G. B. Kauffman and D. O. Cowan, *Inorg. Syn.*, *7*, 239–245 (1963)) are used.—G.B.K.]

[17] [$Na_2[PtCl_4]$ crystallizes with four molecules of water of crystallization. The anhydrous salt may be obtained by heating the tetrahydrate to 150° (W. Peters, *Ber.*, *42*, 4826–4836 (1909), p. 4831; *Z. anorg. Chem.*, *77*, 137–190 (1912), p. 168). $(NH_4)_2[PtCl_4]$ crystallizes without any water of crystallization. It was first analyzed by M. Peyrone (*Ann.*, *55*, 207 (1845)).—G.B.K.]

[18] [$2H_2[PtCl_4] + 8NH_3 \rightarrow [Pt(NH_3)_4][PtCl_4]\downarrow + 4NH_4Cl$.—G.B.K.]

[19] [$[Pt(NH_3)_4][PtCl_4] + 2Na_2CO_3 \rightarrow 2Pt + 4NaCl + 4NH_3\uparrow + 2CO_2\uparrow + O_2\uparrow$.—G.B.K.]

to be $\text{PtCl}^2 + \overline{\text{NH}}^3$ [20] and if the composition is calculated accordingly, one obtains:

Platinum	65.24	[%]
Chlorine	23.42	[%]
Ammonia	11.34[21]	[%]

The green salt does not dissolve either in water, in alcohol, or in hydrochloric acid, nor is it decomposed by the last substance.[22]

---

[20] [Berzelius had introduced the use of barred symbols to designate two atoms (*Jahresbericht*, *7*, 72 (1828)). Thus the formula denotes $PtCl_2(NH_3)_2$. The true composition is a dimer (in a formal sense only) of this formula, viz., $[Pt(NH_3)_4]$-$[PtCl_4]$, tetraammineplatinum(II) tetrachloroplatinate(II).—G.B.K.]

[21] [According to the latest atomic weights, the composition should be Pt 65.02%, Cl 23.62%, and NH_3 11.34%.—G.B.K.]

[22] [According to E. G. Cox, F. W. Pinkard, W. Wardlaw, and G. H. Preston (*J. Chem. Soc.*, *1932*, 2527–2535, p. 2534), the salt may be recrystallized from hydrochloric acid solution.—G.B.K.]

INTRODUCTION TO PAPER TWO

[Although the first organometallic compound was prepared by William Christoffer Zeise in 1827, except for the work of Edward Frankland on metal alkyls, there was little further progress in the field until the work of Kharasch and Anderson in the 1930's.[1] During the past two decades, however, transition metal-olefin complexes have become the subject of intensive research activity.[2] These studies have resulted not only in the discovery of many new compounds but also in the elucidation of their structures and in application of their unusual properties to practical problems. Olefin complexes of the platinum group metals have been recognized as intermediates in a number of homogeneous catalytic processes. Among the reactions undergone by coordinated olefins are isomerization, oxidation, hydrogenation, and polymerization.[3]

Zeise's preliminary report appeared in 1827[4] and hence antedates Paper One of this volume, but his more detailed report,[5] from which Paper Two is excerpted, appeared in 1831. Ethylene ("olefiant gas") was known since 1796[6] or even earlier, but Zeise did not prepare his compound by reaction of $PtCl_2$ with C_2H_4. By refluxing a mixture of platinum(II) and platinum(IV) chlorides in ethanol until the solid had turned black, evaporating off the excess solvent, and treating the residue with potassium chloride, he obtained lemon-yellow crystals of potassium trichloro-(ethylene)platinate(II) monohydrate, $K[Pt(C_2H_4)Cl_3] \cdot H_2O$. Zeise also prepared the ammonium analogue of this compound, but only the potassium compound is known as Zeise's Salt. The synthesis is a complicated process, but since Zeise[7] demonstrated the presence of an aldehyde in his distillate, the reactions have been written:[8]

---

[1] J. S. Anderson, *J. Chem. Soc.*, *1934*, 971–974; *ibid.*, *1936*, 1042–1049.

[2] R. N. Keller, *Chem. Rev.*, *28*, 229–267 (1941); B. E. Douglas, "Coordination Compounds of Metal Ions with Olefins and Olefin-Like Substances," in J. C. Bailar, Jr., ed., *Chemistry of the Coordination Compounds*, Reinhold Publ. Corp., New York, 1956, Chap. 15; R. G. Guy and B. L. Shaw, *Adv. Inorg. Chem. and Radiochem.*, *4*, 77–131 (1962), p. 78; M. A. Bennett, *Chem. Rev.*, *62*, 611–652 (1962); E. O. Fischer and H. Werner, *Angew. Chem. Intern. Ed.*, *2*, 80 (1963); H. W. Quinn and J. H. Tsai, *Adv. Inorg. Chem. and Radiochem.*, *12*, 217–373 (1969).

[3] G. C. Bond, *Platinum Metals Rev.*, *8*, 92–98 (1964).

[4] W. C. Zeise, *Danske Vidensk. Selskabs Overs.*, *1825–1826*, 13; *Pogg. Ann. Phys.*, *9*, 632 (1827).

[5] W. C. Zeise, *Pogg. Ann. Phys.*, *21*, 497–541, 542–549 (1831).

[6] A. F. de Fourcroy, *Acad. des Sciences*, *Mém. divers Sçavans*, *1805*, i, 101.

[7] W. C. Zeise, *Pogg. Ann. Phys.*, *40*, 250 (1837).

[8] K. Birnbaum, *Ann.*, *145*, 67 (1868).

$$PtCl_4 + 2C_2H_5OH \longrightarrow PtCl_2 \cdot C_2H_4 + CH_3CHO + H_2O + 2HCl$$
$$PtCl_2 \cdot C_2H_4 + KCl \longrightarrow KCl \cdot PtCl_2 \cdot C_2H_4$$

Zeise also prepared the related orange compound, *sym-trans*-di-μ-chloro-dichlorobis(ethylene)diplatinum(II), known as Zeise's Dimer,

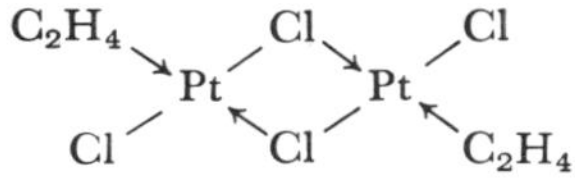

which, along with Zeise's Salt, played an important role in both inorganic and organic bonding theory.[9,10] The two compounds are interconvertible.

Zeise's Salt attracted considerable attention since Zeise formulated it as a derivative of ethylene, even though it was produced from alcohol. At the time, the conflict between the adherents of the Etherin Theory, e.g., Jean Baptiste André Dumas (1800–1884) and Pierre François Guillaume Boullay (1777–1858), and those of the Radical Theory, e.g., Justus von Liebig (1802–1873), was raging. Liebig[11] attacked Zeise's analyses, which indicated the presence of etherin, C_4H_8 (modern C_2H_4), and proposed instead that the compound contained ether, $C_4H_{10}O$, i.e., $(C_2H_5)_2O$. Zeise, however, repeated and confirmed his original analyses.[12] Independent confirmatory proof of Zeise's formulation was furnished in 1868 when Birnbaum[8] prepared Zeise's Salt by passing ethylene gas into a solution of $H_2[PtCl_6]$ and adding KCl.

As time passed, additional related compounds were prepared, and it was recognized that Zeise's Salt was only one of a group of platinum-olefin species. The compound thus stimulated research into other platinum derivatives, which aided in the formulation of a general theory of bonding for metal complexes.[9] The structure and bonding in Zeise's Salt and Zeise's Dimer were better understood after 1951, when ferrocene was prepared and characterized. In 1953 Chatt and Duncanson[13] proposed that the Pt—C linkage consists of two components—(1) one formed by overlap of a Pt dsp^2 hybrid orbital with the C_2H_4 π molecular orbital, thus removing electron density from ethylene, and (2) one formed by overlap of the C_2H_4 empty π^* molecular orbital with the filled Pt 5d orbitals, thus removing electron density from the metal ("backbonding"). The two interactions are mutually reinforcing and strengthen the Pt—C bond while weakening the C═C bond. Theory predicted that the axis of the

[9] J. S. Thayer, *J. Chem. Educ.*, *46*, 442–443 (1969).
[10] J. Chatt and M. L. Searle, *Inorg. Syn.*, *5*, 210–215 (1957).
[11] J. Liebig, *Ann. Chem.*, *9*, 1 (1834); *ibid.*, *23*, 12 (1837).
[12] W. C. Zeise, *Pogg. Ann. Phys.*, *40*, 234 (1837).
[13] J. Chatt and L. A. Duncanson, *J. Chem. Soc.*, *1953*, 2939–2947.

C=C bond should be perpendicular to the plane of the —$PtCl_3$ group:

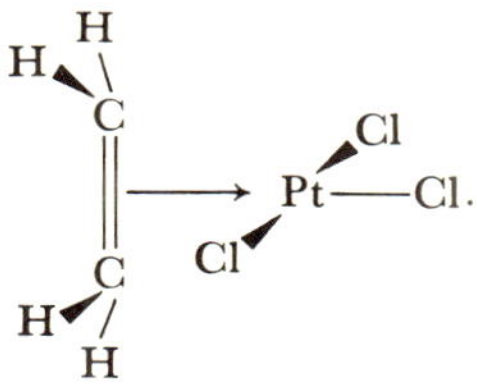

The structure has been confirmed by X-ray diffraction studies and normal coordinate analysis.[14] Thus Zeise's compounds and related transition metal-olefin complexes have provided tremendous insight into non-classical chemical bonding and structures which have no counterpart elsewhere in chemistry.

William Christoffer Zeise[15] was born on October 15, 1789, in Slagelse, Seeland, Denmark, later the birthplace of Sophus Mads Jørgensen (see Paper Six). The son of a pharmacist, Zeise was first an apothecary's apprentice in Sonderburg. Intending to study chemistry, he moved to Copenhagen where he lodged in the home of Hans Christian Ørsted, the physicist and chemist, whose father was a friend of Zeise's father. The two scientists worked in the same laboratory, and together they were to place chemistry in Denmark on a firm basis. Beginning in 1805, Zeise studied pharmacy under Chief Court Pharmacist Becker, who was a Professor at the University of Copenhagen. In 1807 Zeise entered the university and in 1815 passed the pharmaceutical examination. In 1816 he received a traveling fellowship but postponed his trip until after he received his doctorate in 1817, with a dissertation on the effect of alkali on organic materials.[16]

From 1818 to 1820 Zeise traveled to Germany and also Paris where he met Berzelius. On his return he became Ørsted's assistant until 1822, when Zeise became *Extraordinarius Professor* of chemistry at the University of Copenhagen. This marked the first time that pure chemistry had been acknowledged in Denmark as an academic subject. In 1824 Zeise

[14] J. A. Wunderlich and D. P. Mellor, *Acta Cryst.*, *7*, 130 (1954); *ibid.*, *8*, 57 (1955); S. Maričić, C. R. Redpath, and J. A. S. Smith, *J. Chem. Soc.*, *1963*, 4905–4913; M. J. Grogan and K. Nakamoto, *J. Am. Chem. Soc.*, *88*, 5454–5460 (1966); *ibid.*, *90*, 918–922 (1968).

[15] E. Biilman, *Ingeniøren*, *1916*, No. 11, 83–86; J. R. Partington, *A History of Chemistry*, St. Martin's Press, New York, 1964, Vol. IV, p. 352; S. Veibel, *Kemien i Danmark*, Vol. I, Nyt Nordisk Forlag, Arnold Busck, Copenhagen, 1939, p. 155 ff. For a bibliography of Zeise's publications see Vol. II, 1943, pp. 488–494.

[16] W. C. Zeise, *De vi corporum alcalinorum materias regno organico peculiares transmutandi: annexis quibusdam de transmutatione chemica in genere*, Brünnich, Copenhagen, 1817.

became a member of the Royal Danish Society of Sciences, and many of his scientific works were published in the society's transactions. From 1829 on, he held the position of Instructor at the Polytechnic School in addition to his professorship at the university. He died at Copenhagen on November 12, 1847.

Zeise may be regarded as Denmark's first actual organic chemist, but his works encompass both organic and inorganic chemistry, pure and applied, with emphasis on the compounds of sulfur. He was a pioneer in organic analysis and devised a method for the determination of sulfur in organic substances that is still used in principle today.[17] In 1822, by allowing carbon disulfide to react with alcoholic potash, he discovered xanthates,[18] later used in the conversion of cellulose to artificial silk, viscose, and other products. Two years later, by the action of carbon disulfide on ammonia, he discovered thiocarbamic acid.[19] His most famous discovery (1833) is that of the organic sulfur compounds which he named mercaptans (from *corpus mercurio captans*).[20] Despite historical reports to the contrary, it is clear that Zeise, and not Liebig or Berzelius, was the first to recognize the analogy between his new compounds (RSH) and alcohols (ROH). The addition compound of ethylene and platinous chloride, described in the following article (Paper Two), constitutes his most important contribution to inorganic chemistry.—G.B.K.]

---

[17] W. C. Zeise, *J. prakt. Chem.*, *1*, 458 (1834).

[18] W. C. Zeise, *Ann. chim.*, *21*, 160 (1822); *ibid.*, *26*, 66, 113 (1824); *K. Danske Vidensk. Selskabs Skr.*, *1*, 219 (1824).

[19] W. C. Zeise, *Allgem. J. Chem.*, *41*, 98, 170 (1824).

[20] W. C. Zeise, *Ann. Phys.*, *31*, 369 (1834); *J. prakt. Chem.*, *1*, 186 (1834); P. Diergart, *J. prakt. Chem.*, *99*, 281–292 (1919); O. Zeise, *J. prakt. Chem.*, *100*, 48 (1919).

HEINRICH GUSTAV MAGNUS (1802–1870)

[A pencil sketch by Magnus' brother, Eduard. Originally appeared in *Zur Erinnerung an vorangegangene Freunde* by August Wilhelm Hofmann. Braunschweig: F. Vieweg & Sohn, 1888. From *Deutsche Chemiker in der ersten Hälfte des neunzehnten, Jahrhunderts* by Wilhelm Prandtl, p. 303, Weinheim/Bergstrasse, 1956. Courtesy of the publisher, Verlag Chemie, Weinheim/Bergstrasse, German Federal Republic.]

WILLIAM CHRISTOFFER ZEISE (1789–1847)
[Photograph of a painting from *Journal für praktische Chemie*, *99*, 281 (1919). Courtesy of Fräulein Charlotte Werner, Zürich, Switzerland.]

2

William Christoffer Zeise

The Reaction between Platinum Chloride and Alcohol and on the New Substances Formed Thereby

[Translation of excerpts from "Von der Wirkung zwischen Platinchlorid und Alkohol, und von den dabei entstehenden neuen Substanzen," Pogendorff's *Annalen der Physik und Chemie*, *21*, 497–541 (1831); pp. 497–502, 512–519, 532–539 only translated.]

(Translated freely and with some abridgements from the paper transmitted by the author: *De chlorido platinae et alcohole vini sese invicem permutantibus nec non de novis substantiis inde oriundis commentatio, qua festo academico mense novembri A. 1830 celebrato prolusit Dr. W. C. Zeise etc.*)

Considering how vigorously chlorine acts upon alcohol and how weak its affinity is toward platinum, it is certainly not surprising that the chloride of platinum [*Chlorplatin*] also resembles chlorine to some extent in this behavior. The reason for, and the result of, this action, however, is not so easy to discover. For the experiments which I shall report show that in this case almost no ether is formed but instead a new substance consisting of platinum, chlorine, carbon, and hydrogen.

Earlier experiments on the reaction between alcohol and metal chlorides were directed solely toward learning whether ether is obtained; it is therefore possible that we will later discover a similar mode of action among other chlorides.

Five years ago I performed and published in the papers of the Royal Society of Copenhagen for 1825 and 1826[1] some experiments

[1] [W. C. Zeise, *Danske Vidensk. Selskabs Overs.*, *1825–1826*, 13.—G.B.K.]

which opened the way for this investigation.[2] These experiments, however, deal almost exclusively with the peculiar black powder which is formed by the action of platinous chloride [*Platinchlorür*][3] on alcohol. During further investigation of other related phenomena, especially on using platinic chloride [*Chlorid*][4] instead of platinous chloride [*Chlorür*], I later noticed several facts that were worth investigating, some of which I reported long ago to the Royal Society. Detained by other affairs and studies, however, I have only recently been able to complete this work to some degree.

Principal Phenomena on Reacting Platinum Chloride and Alcohol

On moderate warming, platinous chloride which is completely free of platinic chloride changes the alcohol either not at all or to a very slight degree. However, if it is boiled in a distillation flask with alcohol of specific gravity 0.823, the originally green-gray powder gradually becomes black, and the liquid becomes very acidic and smells of ether. But there is so little ether present that not a trace of it can be separated by addition of water, either from the distillate or from the residual liquid. The [supernatant] liquid above the black powder is almost completely colorless.

The phenomena are different if one uses platinous chloride which, because of incomplete calcination, still contains some platinic chloride. For as soon as the alcohol is added, reaction is indicated by an easily noticeable odor of ether, and after three or four days, without application of heat, the chloride is changed into a black powder. The supernatant liquid, however, has meanwhile become brown. If, after the powder has been separated, this liquid is distilled to half its volume, another large amount of that black powder is precipitated.

If platinic chloride, prepared in the usual way, is dissolved in a tenfold amount by weight of alcohol of the specified concentration,

[2] A short report of these experiments is found in these *Annalen*, 9, 632 [(1827)]. P. [Johann Christian Poggendorff (1796–1877), Professor of Physics at the University of Berlin and editor of the *Annalen der Physik und Chemie*. The report is taken from *Årsberättelse om Framstegen i Physik och Chemi*, 7, 129, i.e., Berzelius' *Jahresbericht* in the original Swedish version.—G.B.K.]

[3] [Platinum(II) chloride, $PtCl_2$.—G.B.K.]

[4] [Platinum(IV) chloride, $PtCl_4$.—G.B.K.]

the filtered liquid reduced to one-sixth its volume over a small flame, and the intensely colored residue, which still contains much platinum, is tested with ammonium chloride, then no further trace of platinic chloride is found in it, and although sometimes more, sometimes less of the black powder is precipitated, ether can be detected only by its odor. However, it should be noted that a considerable amount of free hydrochloric acid is present in the distillate as well as in the residual liquid.

If an appropriate amount of potassium chloride is added to the substance and the mixture is evaporated, there is formed the flammable salt which I shall hereafter call *flammable potash-platinum salt* [*entzündliches Kali-Platin-Salz*].[5] It forms very beautiful, yellow crystals. A similar salt is obtained if ammonium chloride is used instead of potassium chloride.[6] If the brown liquid is allowed to evaporate, it leaves behind an amorphous, brown mass, in which, however, yellow and black particles may be distinguished. On stronger heating this mass swells violently, expelling a flammable vapor and hydrogen chloride; finally, when the vessel has not yet begun to glow, the mass starts to burn, and it leaves behind spongy platinum of a silver-white color. For simplicity, I will call this mass *crude flammable chloride* [*rohes entzündliches Chlorid*], and the liquid, *crude liquid* [*rohe Flüssigkeit*].

If this crude flammable chloride, prepared by careful evaporation, is macerated with water, quite a bit of it dissolves, but a brown slimy mass always remains behind. No matter how often this platinum deposit is washed, it contains much chlorine and also carbon and hydrogen. For if it is heated in a glass dish, it expels hydrogen chloride, and finally, on increased heating, it leaves behind platinum. When warmed with *aqua regia*,[7] a considerable amount of it dissolves, while another brown substance, which floats like molten resin in the liquid, resists this solvent. If this mass is examined after digestion has been completed, it appears brittle and flammable; when heated in a glass dish, it swells exceedingly, then burns, and leaves behind spongy platinum.

---

[5] [$H[Pt(C_2H_4)Cl_3] + KCl + H_2O \rightarrow K[Pt(C_2H_4)Cl_3]\cdot H_2O$ (Zeise's Salt) + HCl. Zeise called the salt *sal kalico-platinicus inflammabilis* in his Latin paper of 1830.—G.B.K.]

[6] [$H[Pt(C_2H_4)Cl_3] + NH_4Cl + xH_2O \rightarrow NH_4[Pt(C_2H_4)Cl_3]\cdot xH_2O + HCl$.—G.B.K.]

[7] [A solvent mixture dating from alchemical times and consisting of three or four volumes of conc. HCl to one volume of conc. HNO_3.—G.B.K.]

This residue of the crude platinic chloride is undoubtedly formed partially from the pure chloride decomposed by the evaporation; partially, however, it seems to have been previously present in the crude liquid dissolved by the hydrochloric acid.

If the yellow-brown liquid from the maceration of the crude chloride is again evaporated to dryness, the residue exhibits a yellow or yellow-brown color, especially if it is freed from the admixed, very soluble, brown substance by washing it with some water. In a greater amount of water, this mass dissolves with a yellow color, but brown particles always remain behind. If this solution is evaporated again *in vacuo* over caustic potash,[8] the substance which I shall henceforth describe under the name *flammable platinum chloride* [*entzündliches Platinchlorid*][9,39] is produced, as pure as is possible by this method.

If, from the beginning, the crude liquid is evaporated *in vacuo* without heating, a greater amount of the pure chloride can be extracted, but here too on evaporation much of it is converted into that brown insoluble substance.

Before I turn to the further investigation of this compound, I must say something about the action of the chloride on alcohol. All the experiments which I performed regarding this point indicate that before the chloride disappears completely, some of that black deposit is formed; but I do not consider its production to be essential for the decomposition of the chloride, for its amount varies according to the circumstances, very small in a platinic chloride completely free of platinous chloride but larger in a platinic chloride contaminated with it.

Platinic chloride, which is concentrated in the usual way until no further hydrogen chloride is given off, always contains much platinous chloride, which for the most part is dissolved not only by the water but also by the alcohol. Before the complete conversion of such a platinic chloride, a fourth or sometimes a third of it is

[8] [Potassium hydroxide, KOH.—G.B.K.]

[9] [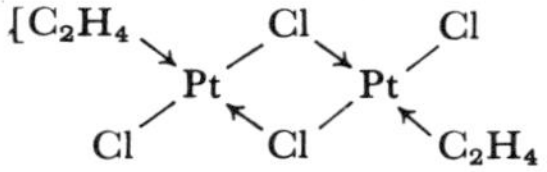

(Zeise's Dimer). For its synthesis as well as that of Zeise's Salt see J. Chatt and M. L. Searle, *Inorg. Syn.*, *5*, 210–215 (1957). In his Latin paper of 1830 Zeise called this compound *chloridum platinae inflammabile*.—G.B.K.]

deposited as that black powder. However, if chloride which has been evaporated *in vacuo* finally over caustic potash[8] is used, not even a hundredth of it is converted into that deposit. But in order for this to occur, one should not use more alcohol than necessary and on distilling the liquid a certain concentration should not be exceeded, for the flammable chloride itself is destroyed by alcohol so that it causes that deposit to precipitate.

Thus it is probable that that deposit is formed either from the admixed platinous chloride or from the decomposition of the flammable chloride by the excess alcohol.

Although the proportion of alcohol cited above is usually sufficient for this operation, a larger amount is sometimes necessary. In some cases the platinic chloride was so resistant that it had to be boiled three to four times with the specified amount of fresh alcohol. Then sometimes, but not always, the decomposition was accelerated by addition of water. Indeed, I have not been able to ascertain satisfactorily the cause of this resistance, but I have to note that it cannot be attributed solely to the hydrochloric acid, or the nitric acid, or to the varying amount of water.[10]

Flammable Platinum-Potash Salt [Entzündliches Platin-Kali-Salz][5]

It is known that potassium chloride, like ammonium chloride, yields a water-soluble compound with platinous chloride. Both compounds, which were carefully investigated several years ago by G. Magnus [1802–1870],[11] form red crystals, which are completely insoluble or very difficult to dissolve in alcohol.[12] Therefore, when in the experiments described on p. 23, yellow crystals readily soluble in alcohol were produced, it was easy to see that the flammable platinum chloride had combined unchanged with the potassium chloride.[13,14]

---

[10] [The sections *Fulminating Platinum Deposit* [*Verpuffender Platinabsatz*] and *Flammable Platinum Chloride* [*Entzündliches Platinchlorid*], pp. 502–512 of the original article, have been omitted in this translation.—G.B.K.]

[11] These *Annalen, 14*, 239[–242 (1828). See Paper One of this volume.—G.B.K.]

[12] [Potassium and ammonium tetrachloroplatinates(II), $K_2[PtCl_4]$ and $(NH_4)_2$-$[PtCl_4]$.—G.B.K.]

[13] See Berzelius' *Jahresbericht, 9*, 162 [(1830)] or these *Annalen, 16*, 82 [(1824)].

[14] [$\underset{\text{Zeise's Dimer}}{[Pt_2(C_2H_4)_2Cl_4]} + 2KCl + H_2O \rightarrow \underset{\text{Zeise's Salt}}{2K[Pt(C_2H_4)Cl_3]\cdot H_2O}$.—G.B.K.]

Under favorable conditions, the flammable platinum-potash salt forms very beautiful crystals,[15] often one-half inch or more in length. They are lemon-colored and almost completely transparent. In the course of time, however, especially upon introduction of air and light, they gradually become coated with a black, opaque crust. In very dry air they also lose some water. *In vacuo* over sulfuric acid they soon become opaque, but only after pulverization do they lose all their water of crystallization.

Near the boiling point of water they lose to the air the same amount of water as on prolonged storage *in vacuo*.

The taste of this salt is metallic, astringent, and persistent. It reddens litmus paper. Its solutions are yellow. On moderate heating, it requires five times its weight of water and somewhat more of alcohol in order to dissolve.

When heated in an open vessel, it soon becomes black (probably at no more than 200°); on more intense heating, hydrogen chloride gas appears, mixed with flammable gas (as with flammable platinum chloride itself), and finally, after combustion has occurred, the entire material is converted into a gray mass. If anhydrous salt is used in this process, no effervescence is noted. From the gray residue, water extracts not only potassium chloride but also some platinum-potash salt, even if very intense heat has been used, so that decomposition by mere ignition is useless for the analysis of this salt.

When I exposed this salt in a pneumatic distillation apparatus to the heat of boiling water, as long as water still distilled over, I could not afterwards, on stronger heating, detect any further trace of water. But it is very difficult to dry the salt this way in a retort. Indeed, often when I performed the experiment with different portions of the salt which I believed to be well dried, a large amount of moisture appeared at exactly the same moment when, upon increased heating, the chemical decomposition of the mass began, so that I do not doubt that a certain portion of the water is inherent in the salt in such a manner that it can be expelled only when decomposition has already begun. For if the previously dried salt is not quickly poured into a retort, exposed to the heat of boiling water for

[15] Prof. [Johann] G. Forchhammer [(1794–1865), Lecturer (1823) and Professor (1835) of Chemistry and Mineralogy at the University of Copenhagen], who investigated the crystal form at my request, found it to be hemiprismatic. The inclination between the prismatic faces amounted to 103° 58′ and that between the prismatic faces and the terminal face 112° 5′. The greatest deviation in the first measurement amounted to 10′.

a long time, and if the resulting coating of moisture on the glass is not repeatedly removed, moisture always appears whenever more intense heat is applied, whereby the mass becomes black. If I am not mistaken, this seems to me to prove a remarkable attraction of the dry salt for water. It should also be noted here that no considerable trace of ether appears among the decomposition products of the salt if the heat is increased only after complete expulsion of the water; otherwise the odor of ether becomes perceptible.

Perhaps it is not superfluous to add that the mass, if it is decomposed with exclusion of air, does not ignite, and the evolution of gas (hydrogen chloride and hydrocarbon) always ceases before the temperature has reached a red heat.

When dissolved in water, the flammable platinum-potash salt undergoes the same change as the flammable platinum chloride alone, except that on moderate heating it is destroyed less rapidly. It is remarkable that although this salt in the solid state can withstand a temperature of 100° whereupon it produces only water, yet when dissolved in water it already decomposes into its constituents at about 90°. However, some platinum always remains dissolved.

Above all, the action of hydrochloric acid is quite remarkable here. For if a certain amount of this acid is added to the solution of the flammable salt, no sign of decomposition is visible, even after prolonged boiling. For this reason it also happens that the crystals of this salt maintain their color and luster for a long time, as soon as they are contaminated with hydrochloric acid.

Sulfuric acid or nitric acid exerts almost the same effect upon the solution, but it should be noted that with nitric acid no ordinary platinum-potash salt ($PtCl^4 + KCl^2$) [16] is formed.

Ample addition of potassium chloride sometimes seems to retard the decomposition.

Caustic potash,[8] added to the solution of the flammable platinum-potash salt, seems initially to cause no change, but after about 20 minutes that brown, slimy mass (p. 508) [10] is deposited, and the liquid becomes almost completely colorless. If, after the addition of the potash, the liquid is immediately heated, one notes exactly the same phenomena as with the flammable platinum chloride (p. 509),[10] so that that exploding precipitate can also be prepared in this way.

---

[16] [Potassium hexachloroplatinate(IV), $K_2[PtCl_6]$.—G.B.K.]

Ammonia water exerts a peculiar effect here, for as soon as some of this is poured into the solution of the flammable platinum-potash salt, a voluminous yellow precipitate soon forms. Because of lack of time and lack of a sufficient amount of this substance I have not yet investigated it properly. I have, however, convinced myself that it is somewhat soluble in water, much more in hot than in cold, that it is decomposed over a flame, and that after sufficient heating in an open vessel it leaves behind much platinum.[17]

I have investigated the nature of this salt by means of silver nitrate in the same way which was explained for flammable platinum chloride on p. 509,[10] and in this case I have observed the same phenomena and the same peculiar result.[18]

The salt is attacked by hydrogen gas, even at room temperature, and by wet as well as by dry methods. If this gas is conducted through a solution of the flammable platinum-potash salt, platinum is soon precipitated, and if the operation is continued for a long time, most of this metal is precipitated, but some of it always remains [behind in solution]. If the gas is conducted over the salt after it has been powdered, dried over boiling water, and placed in a horizontal glass tube, then very slight traces of moisture are usually noted at the moment when the salt is attacked and blackened by the gas with development of heat. The complete decomposition of the salt, however, can be accomplished only on heating.[19]

Chlorine gas does not react with the dry salt at room temperature. However, if the salt is subjected to the action of chlorine, aided by heating in a glass tube which is provided with a receiver and with delivery tubes, a peculiar phenomenon occurs. For not only is hydrochloric acid thus formed but also some moisture is formed in the

---

[17] A more detailed investigation of this substance [*trans*-$[Pt(NH_3)(C_2H_4)Cl_2]$ is found in the following article. [This paper, "Gekohlenwasserstoffes Chlorplatin-Ammoniak," may be found in *Pogg. Ann.*, *21*, 542–549 (1831).—G.B.K.]

[18] [In this untranslated section, Zeise found that treatment of $[Pt_2(C_2H_4)_2Cl_4]$ with silver nitrate produced an abundant precipitate. The filtrate decomposed quickly with precipitation of all the platinum, while the filtrate from this second precipitation gives on warming a further precipitate of silver chloride. From this Zeise concluded "*that in this compound a portion of the chlorine is present in the same manner as in metal chlorides, however, another portion in the manner as in ethylene chloride* (chloric ether, *Chloräther*) *and ethyl chloride* [*Chlorwasserstoffäther*], compounds in which the chlorine content can be discovered only after their destruction. Accordingly, it is very probable *that the flammable chloride is a compound of platinum chloride and* [a] *chlorinated hydrocarbon*" (Zeise's italics).—G.B.K.]

[19] [Possibly, $K[Pt(C_2H_4)Cl_3]\cdot H_2O + 2H_2 \rightarrow Pt + KCl + 2HCl + C_2H_6 + H_2O$.—G.B.K.]

receiver, and in addition there are obtained colorless, volatile, and aromatic-smelling crystals, which are undoubtedly nothing but carbon tetrachloride [*Chlorkohlenstoff*].[20] For, as is well known, carbon tetrachloride possesses this property, and it is certain that this is formed when chlorine reacts with chloric ether [*Chloräther*],[21] that is, with the compound of chlorine, hydrogen, and carbon.

If chlorine is conducted through the solution of the flammable salt, an odor of ether is perceptible.

Iodine and bromine act similarly upon the solution of the salt.

Now I need to speak in more detail of the preparation of the platinum-potash salt, which, according to my experience, is best accomplished in the following manner.

The crude liquid [*rohe Flüssigkeit*], which must be properly prepared so that it is no longer made turbid by concentrated potassium chloride solution,[22] is decanted from the precipitate after dilution with about four times its weight of water and is mixed with powdered potassium chloride so that 1 part of this corresponds to 4 parts of the dry chloride originally used for the crude liquid. As soon as this is dissolved by the liquid, to which more water can be added if necessary, it is evaporated over a moderate flame to about a third of the volume. After cooling, a mass which usually consists of brown crystals is found in the vessel. After this is powdered and the supernatant liquid, which is even browner and very acidic, has been decanted, the crystals are dissolved in water. At this point some brown-red crystals appear. I do not know whether they always contain iridium, for they also appeared during several experiments in which I used a chloride in which I could discover none of this metal.

The new solution is evaporated again to about one third of its volume, but over a very small flame, at least near the end. On cooling, beautiful crystals form. After separation from the liquid, these are washed with very little water and thereupon redissolved. The liquid prepared in this manner, which is yellow and completely

---

[20] [Since CCl_4 melts at $-22.6°C$, this may possibly be hexachloroethane, a polychlorinated ethane.—G.B.K.]

[21] [Ethylidene chloride, 1,1-dichloroethane, "oil of the Dutch Chemists," or simply "Dutch liquid," CH_3CHCl_2.—G.B.K.]

[22] With regard to the preparation of this liquid, it is probably not superfluous to remind the reader that as soon as a great deal of the black powder has deposited, one must proceed carefully since otherwise the mass expands so violently that a great part of it spurts over into the receiver.

odorless, is finally allowed to evaporate *in vacuo* without heating.

Why one must proceed in this manner is easily seen from the facts cited on p. 27. For so much free hydrochloric acid is present in the crude liquid that no decomposition need be feared during the first two evaporations; however, as soon as the acid is removed, heat cannot be used without danger. Indeed, even if the evaporation is performed *in vacuo*, marks of decomposition occasionally appear unless the evaporation is accelerated by agitating the sulfuric acid placed next to it and replacing the first portion of acid with a new one after its drying action has been exhausted.

At times, toward the end of the process, I have also placed caustic potash[8] next to the crystals in order to remove more easily the residual hydrochloric acid, a procedure which is not necessary, however, as long as the evaporation is moderated to such an extent that about a fifth of the liquid still remains, and then the already distinctly formed crystals are washed with very little water. If the crystals are dried afterwards with boiling water, it is not even necessary to proceed so carefully.

From the decanted portions of the liquid, even from the last and the next to the last, a few more crystals can be obtained in this manner, but they cannot be obtained in a pure state without much work. Ordinarily, a substantial amount of potassium chloride crystallizes only from the first portion of the decanted liquid; later brown-red crystals also appear. I have not investigated these further because of their small quantity.

I must note again here, as on p. 23, that in the crude liquid, in addition to the flammable platinum chloride and the hydrochloric acid, there is always present the substance which imparts a brown color and a peculiar odor to that liquid. But whether this substance is formed simultaneously with the flammable platinum chloride or is a result of its decomposition is still undecided.

We now come to the more difficult part of the investigation, namely, to the analysis of this flammable salt, which I have performed with great care; for the precise knowledge of the components of this substance seemed very desirable to me, not only because this compound represents a union of elements which very rarely occur together,[23] but also because an accurate determination of the ratios

[23] [That is, platinum and carbon. However, today many organometallic compounds are known (see the Introduction to this paper).—G.B.K.]

between carbon, hydrogen, and chlorine might perhaps throw some light on the nature of chloric ether[21] and other compounds of this type, in the investigation of which [César Mansuète] Despretz [1792–1863], [Georges Simon] Serullas [1774–1832], [Henry] Hennel [?–1842], [Jean Baptiste André] Dumas [1800–1884] and [Pierre François Guillaume] Boullay [1777–1858], [H.] Morin, and others have recently expended much effort.[24]

In 100 parts of the anhydrous salt was found:

A. By experiments in which only the amount of platinum and the total amount of chlorine were determined (p. 522):[25]

Platinum	51.3326	51.254
Chlorine	28.3119	28.563.

B. By experiments in which the amounts of platinum, potassium chloride, and remaining chlorine were determined (p. 524):[26]

Platinum	50.4535		
Potassium Chloride	19.5661 =	Potassium	10.2785
		Chlorine	9.2876
Remaining Chlorine	18.3610.		

C. By experiments in which only the amounts of platinum and potassium chloride were determined (p. 525):[27]

Platinum	50.198	52.5659
Potassium Chloride	20.375	19.74406.

---

[24] [The analytical details, which comprise pp. 519–532 of the original paper, are not translated here, but the summary and discussion of the analytical results are given. The analytical procedures are described briefly in footnotes 25 to 31.—G.B.K.]

[25] [The sample was ignited in a crucible with potassium carbonate and the residue extracted with water. The remaining platinum was determined by weighing, and the chlorine in the washings was precipitated with silver nitrate and weighed as silver chloride.—G.B.K.]

[26] [The sample was heated in a glass tube in a current of hydrogen, the evolved hydrogen chloride was absorbed in aqueous ammonia, and the chlorine in the resulting solution was precipitated and weighed as silver chloride. The residue in the ignition tube was extracted with water, and the remaining platinum was dried and weighed. The washings were evaporated, and the resulting potassium chloride was weighed.—G.B.K.]

[27] [These analyses were carried out like the ones described in the preceding footnote, but the current of hydrogen and the absorption in ammonia were omitted.—G.B.K.]

D. By experiments in which only the amounts of carbon and hydrogen were determined (pp. 525–529):[28]

Carbon	7.1520	6.94545	6.8861	6.61910
Hydrogen	1.53259	1.64035	1.1987	1.04169.

E. By an experiment in which only the amount of carbon was determined (p. 530):[29]

Carbon 5.7113.

F. By experiments in which only the amount of hydrogen was determined (pp. 531, 532):[30]

Hydrogen 1.4212 1.0531.

According to experiments (p. 520),[31] the amount of water in 100 parts of the crystallized salt corresponded to:

5.324 4.505 4.737 5.215 5.173 6.6127.

On the basis of the atomic weights adopted by [Jöns Jacob] Berzelius [1779–1848],[32] the data yield the following composition for the anhydrous flammable potash-platinum salt:

			In 100.
2 Atoms of Platinum	=	2466.5200	53.157203
4 ,, ,, Chlorine	=	885.3000	19.079542
1 Atom ,, Potassium	=	489.9160	9.539773
2 Atoms ,, Chlorine	=	442.6500	10.558424
4 ,, ,, Carbon	=	305.7440	6.589242
8 ,, ,, Hydrogen	=	49.9184	1.075816
		4640.0484	100.000000.

[28] [The sample was mixed with cupric oxide and heated, a procedure introduced by Johann Wolfgang Döbereiner (1780–1849). The evolved water was absorbed in calcium chloride (a procedure introduced by Berzelius), weighed, and the weight of hydrogen calculated. The weight of carbon was calculated from the volume of evolved carbon dioxide, which was collected over mercury.—G.B.K.]

[29] [This analysis was carried out as described in the preceding footnote, but the calcium chloride drying tube was omitted. Zeise attributed the low value obtained to an error in measurement.—G.B.K.]

[30] [These analyses were carried out by heating the sample, in one experiment alone and in the other mixed with platinous chloride. The evolved gases (hydrogen chloride and hydrocarbon) were collected over mercury.—G.B.K.]

[31] [In these analyses, the powdered or crystalline samples were either allowed to stand *in vacuo* over sulfuric acid or were heated. The loss in weight was attributed to water.—G.B.K.]

[32] [The value of O = 16 as the standard for atomic weights did not come into consistent use until the middle of the nineteenth century. Thomas Thomson

Furthermore, in the crystallized salt two atoms[33] of water equals 224.96 according to which the atomic[33] weight of the crystallized salt becomes 4865.0084 and its water content in hundredths equals 4.625. According to the experiments, excluding the last, the average value for the water content in hundredths corresponds to 4.97.[34]

The deviation of the experimental results from the calculated results is, in my opinion, nowhere so great that we cannot consider the assumed composition to be correct. Of course, some rather large deviations occur here and there; but if we consider that loss of platinum is almost unavoidable in these experiments,[35] that the greater amount of water in the first and second experiments, p. 526 and p. 527,[24] must be attributed to the inadequate method of drying, that in the determination of water of crystallization it is very difficult to free the crystals completely of all adhering water, and finally that the crystals were weighed as a powder, then, if at the same time we recall the data reported for the individual experiments, we will surely find the deviations understandable, even if not excusable.[36]

However, I must still note that the crystals which lost 6.6 per cent water in the experiment described on p. 521[24] were not only larger than the usual ones, as I have already mentioned, but also deviated somewhat from them in form.

---

and William Hyde Wollaston assigned oxygen values of 1 and 10, respectively. The values used here by Zeise are based on Berzelius' table of 1826 (*Jahresbericht über die Fortschritte der physischen Wissenschaften*, 7, 73 (1828)), in which O = 100.—G.B.K.]

[33] [Zeise used the terms *atoms* and *atomic weight* for what we would call *molecules* and *molecular weight*.—G.B.K.]

[34] [Omitting the last value of 6.6127 and averaging the remaining five values gives 5.00.—G.B.K.]

[35] That some platinum has been lost may also be deduced from the fact that, e.g., in the first experiment, p. 524 [(not included in this translation)], the amount of chlorine was larger than it should have been according to the amount of platinum obtained; for 50.4535 platinum requires 18.109 chlorine, whereas 18.361 was found; and still here too a loss of chlorine was unavoidable.

[36] Morin, in his paper on the action of chlorine on ethylene [*Doppelt-Kohlenwasserstoff*], says that the flammable gas, which is formed from the chloric ether[21] destroyed on heating, has a composition such that 1 vol. of it consumes 1.6 vol. of oxygen on combustion and yields 0.6 vol. of carbon dioxide [*Kohlensäure*]. If this is so, it becomes easy for me to explain why some of my experiments yielded too large an amount of carbon; for the larger the amount of admixed flammable gases (of which, as I have assumed, 1 vol. gives 1 vol. of carbon dioxide on combustion), the smaller, then, was the calculated amount of carbon. See p. 527.[24]

It is also noteworthy that the larger loss in weight agrees very well with the taking up of 3 atoms[33] of water of crystallization, for this gives 6.779 per cent. I still have a few of these crystals, which I obtained once during a very slow evaporation, but which I did not wish to use up because of their rarity and beauty.

From the proportions of the constituents of the flammable potash-platinum salt, the following composition results for the *flammable platinum chloride* [*entzündliches Chlorplatin*]:[9,32]

			In 100.
1 Atom of Platinum	=	1233.2600	66.528
2 Atoms ,, Chlorine	=	442.6500	23.879
2 ,, ,, Carbon[37]	=	152.8720	8.247
4 ,, ,, Hydrogen	=	24.9592	1.346
		1853.7412[38]	100.000.

Now there still remains the question of the actual structure of the flammable platinum chloride. *Platinous chloride* [*Platinchlorür*] consists of 1 atom of platinum and 2 atoms of chlorine, *ethylene* [*Doppelt-Kohlenwasserstoff*] of 2 atoms of hydrogen and 1 atom of carbon, and *ethylene chloride* [*Kohlenwasserstoffchlorür*] or the chloric ether,[21] prepared from alcohol, of 2 atoms[33] of *ethylene* [*Kohlenwasserstoff*] and 1 atom of chlorine, according to the usual assumption. It may now be asked whether the formula for the flammable platinum chloride[9] is:[39]

$$PtCl^2 + 2H^2C,$$ [40]

[37] [In the original paper this reads "hydrocarbon" [*Kohlenwasserstoff*] instead of "carbon" [*Kohlenstoff*].—G.B.K.]

[38] [In the original paper the molecular weight is incorrectly given as "1854.9412." The percentages given, however, correspond to the correct molecular weight.—G.B.K.]

[39] In order to prevent any misunderstanding, I must note here that on the first galley sheet (*Bogen*) of this paper, i.e., on p. 510,[10] the word *Platinchlorid*, as in the original Latin text, has been used, often in a general sense, as synonymous with *Chlorplatin* to designate any compound of chlorine and platinum, consequently not as a special designation for the highest *Chlorplatin* ($PtCl^4$). According to the author's analysis, the compound in question, which was earlier called *flammable platinum chloride* [*entzündliches Platinchlorid*] [*Chloridum platinae inflammabile*], is a compound of either *platinous chloride* [*Platinchlorür*] with *ethylene* [*Doppelt-Kohlenwasserstoff*] or of a hitherto unknown *platinous subchloride* [*Platinsubchlorür*] [PtCl] with ethylene chloride [*Chlorkohlenwasserstoff*]. Therefore, if one does not wish to express the degree of chlorination [*Chlorungsstufe*] of the platinum in the name, it is indisputably more correct to say *flammable platinum chloride* [*entzündliches Chlorplatin*].

[40] [In modern formulation, $PtCl_2 + C_2H_4$, but actually $2PtCl_2 + 2C_2H_4$.—G.B.K.]

or:

$$PtCl + (2H^2C + Cl),$$

and the formula for the flammable potash-platinum salt:[5]

$$2PtCl^2 + 4H^2C + KCl^2,$$[41]

or:

$$2PtCl + 2(2H^2C + Cl) + KCl^2,$$

or, if desired:

$$2(PtCl + 2H^2C + Cl) + KCl^2.$$

Above all, we must consider here how the flammable platinum chloride and its compound with potassium chloride behaves toward silver nitrate (p. 510).[18] If it is assumed that the entire chlorine content of the flammable platinum chloride is bound to the platinum, then it cannot be understood at all why only a part of the chlorine goes with the silver. However, if it is assumed that only one atom of chlorine is combined with the platinum, but the other is combined with the ethylene [*Kohlenwasserstoff*], then it is clear why only the platinum but not the ethylene [*Kohlenwasserstoff*] yields its chlorine to the silver, as the experiment has indeed shown.

Now if the formula $PtCl + 2H^2C + Cl$ is assumed here for the flammable platinum chloride, then this substance is seen to be a compound of two chlorides, and the flammable potash-platinum salt a compound of three chlorides (or, if one prefers, a compound of a complex [*zusammengesetzt*] chloride and a simple chloride), and indeed this agrees with what we know of the oxides and sulfides [*Sulfureten*]. To assume that the flammable chloride is a compound of a chloride with a carbide [*Carbonid*] and the flammable salt a compound of two chlorides with a carbide (like the compounds of oxides with sulfides) would not be completely absurd, of course, but it does not seem to correspond too well with the usual behavior of the chemical compounds.

Yet it has not escaped me that, on the other hand, there are phenomena which seem to favor the view that the flammable platinum chloride[9] consists of platinous chloride [*Platinchlorür*] and

---

[41] [In modern formulation, $PtCl_2 + C_2H_4 + KCl$.—G.B.K.]

ethylene [*Doppelt-Kohlenwasserstoff*], as incomplete as were the attempts to separate the ether completely (p. 511).[10] In order to explain the phenomenon with silver nitrate, however, it must then be assumed that a certain amount of the platinous chloride is bound so strongly by the ethylene [*Doppelt-Kohlenwasserstoff*] that it is not decomposed by the silver nitrate.[42]

Probably this is also the place to consider a recent publication by [H.] Morin,[45] "On the Action of Chlorine on Ethylene" [*Über die Wirkung des Chlors auf den Doppelt-Kohlenwasserstoff*], in which he attempts to prove by new experiments: (*a*) that the oily substance which is formed from chlorine and olefiant gas[46] is completely identical with the chloric ether[21] which is formed by the action of chlorine on alcohol or ether; and (*b*) that one as well as the other contains 2 atoms of carbon, 3 atoms of hydrogen, and 1 atom of chlorine. It therefore follows (α) that there is no such compound of this type which contains 2 atoms of carbon to 2 atoms of chlorine (which, however, was claimed earlier for the substance resulting from the olefiant gas);[46] and (β) *that, regardless of how the Chlorkohlenwasserstoff may be prepared, the carbon and hydrogen never maintain the same ratio in it that they do in the ethylene* [*Kohlenwasserstoff*], as was likewise reported earlier. For this reason, the above-mentioned chemist regards the chloric ether[21] as consisting of one atom[33] of carbon tetrachloride [*Chlorkohlenstoff*] (CCl^2)[47] and 3 atoms[33] of ethylene [*Doppelt-Kohlenwasserstoff*] (C^3H^6).[48]

Accordingly, I would also suspect 6 atoms of hydrogen to 4 atoms of carbon in the flammable platinum chloride, if such a ratio did not deviate so much from my experiments. Therefore, if one assumes Morin's statements to be correct, then it seems proven to me

[42] Moreover, it should be noted that, regardless of how its composition is viewed, the flammable potash-platinum salt contains two atoms[33] of platinum chloride [*Chlorplatin*] to 1 atom[33] of potassium chloride, whereas the ordinary potash-platinum salt [*das gewöhnliche Kali-Platin-Salz*][43] = $PtCl^4 + KCl^2$, and the red potash platinum salt[44] [*das rothe Kali-Platin-Salz*], according to Magnus,[11] = $PtCl^2 + KCl^2$.

[43] [Potassium hexachloroplatinate(IV), $K_2[PtCl_6]$. For synthesis see G. B. Kauffman, *Inorg. Syn.*, *9*, 182–185 (1967).—G.B.K.]

[44] [Potassium tetrachloroplatinate(II), $K_2[PtCl_4]$. For synthesis see G. B. Kauffman and D. O. Cowan, *Inorg. Syn.*, *7*, 239–245 (1963).—G.B.K.]

[45] *Ann. chim. phys.*, [[2]] *43*, 325 [(1830)], and these *Annalen*, *19*, 61 [(1830)].

[46] [Ethylene, C_2H_4.—G.B.K.]

[47] [In modern formulation, CCl_4.—G.B.K.]

[48] [In modern formulation, $3C_2H_4$.—G.B.K.]

that chlorine in its compound with platinum exerts a different decomposing action on alcohol than it does alone.[49]

Now if one assumes with Morin that ordinary chloric ether[21] consists not of chlorine and a certain hydrocarbon, that is, H^3C^2, but rather of carbon tetrachloride [*Chlorkohlenstoff*] and ethylene [*Doppelt-Kohlenwasserstoff*], then one may regard the flammable platinum chloride[9] as analogous to this compound, if one expresses oneself thus (neglecting the behavior of the silver nitrate), *the platinum in the flammable platinum chloride*[9] *is, so to speak, the substitute of that portion of the carbon which is bound to the chlorine in the ether*; so that, if the *chloric ether*[21] is a compound of *one* atom[33] of *Kohlenchlorür* and three atoms[33] of *ethylene* [*Doppelt-Kohlenwasserstoff*], the flammable platinum chloride,[9] on the other hand, would be a compound of *one* atom[33] of platinous chloride [*Platinchlorür*] and *two* atoms[33] of ethylene [*Doppelt-Kohlenwasserstoff*].[40]

University of Copenhagen

[49] One will perhaps ask what becomes of the carbon of the alcohol, which is separated out by the liberation of hydrochloric acid [*Chlorwasserstoffsäure*]. One should, however, only remember the substance which colors the red liquid brown and imparts to it a peculiar odor, p. 519.

THOMAS GRAHAM (1805–1869)
[From *Journal of Chemical Education*, *11*, 281 (1934).]

INTRODUCTION TO PAPER THREE

[As the number of known coordination compounds increased, theories to explain their constitution were devised. The Scottish chemist Thomas Graham (1805–1869) is credited with originating the first theory of metal-ammines, the so-called ammonium theory, in which metal-ammines are considered as substituted ammonium compounds. In 1837 Graham[1] attempted to explain the constitution of compounds such as diammine-copper(II) chloride by formulas such as

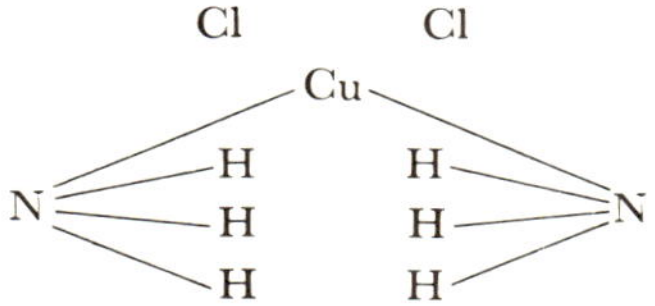

Because of the close analogy between copper and hydrogen, he suggested that two hydrogen atoms, one from each ammonia molecule, had been displaced by the copper atom. In other words, he viewed the salt as a "chloride of cuprammonium" similar to the "chloride of ammonium." He thus accounted for the strong retention of the ammonia in the complex. The theory shows a close affinity to the modern Lewis acid-base approach to the phenomenon of coordination. Despite the fact that Graham's ammonium theory could be applied only when the number of ammonia molecules in the coordination compound is equal to the electrovalence of the metal, it met with a fair degree of success and was generally accepted until Werner's time, largely because of the modifications of it that were proposed by other chemists such as Gerhardt,[2] Wurtz,[3] Reiset,[4] Hofmann,[5] Boedecker,[6] and Weltzien.[7]

Charles Frédéric Gerhardt (1816–1856)[2] applied Graham's idea to the platinum-ammines, which he regarded as ammonia molecules in which different amounts of hydrogen had been replaced by platinum. He considered Reiset's first and second chlorides as compounds of hydrogen

---

[1] T. Graham, *Elements of Chemistry* [J. B. Baillière?], London, 1837. This book is rare and is better known through its German translation, *Lehrbuch der Chemie*, transl. by F. J. Otto, F. Vieweg und Sohn, Braunschweig, 1840, Vol. 2, p. 741.

[2] C. Gerhardt, *Jahresbericht*, *3*, 335 (1850).

[3] C. A. Wurtz, *Ann. chim. phys.*, [3] *30*, 488 (1850).

[4] J. Reiset, *Compt. rend.*, *18*, 1100 (1844); *Ann. chim. phys.*, [3] *11*, 417 (1844).

[5] A. W. von Hofmann, *Ann.*, *78*, 253 (1851).

[6] C. H. D. Boedecker, *Ann.*, *123*, 56 (1862).

[7] C. Weltzien, *Ann.*, *97*, 19 (1856).

chloride with ammonia in which hydrogen had been replaced by *platinosum* (divalent platinum):

Reiset's First Chloride (modern $[Pt(NH_3)_4]Cl_2$):

$$PtCl + 2NH_3 = N_2H_5Pt + HCl$$

Reiset's Second Chloride (modern *trans*-$[Pt(NH_3)_2Cl_2]$):

$$PtCl + NH_3 = NH_2Pt + HCl.$$

Gerhardt designated NH_2Pt as *platosamine* and N_2H_5Pt as *diplatosamine.* Charles-Adolphe Wurtz (1817–1884)[3] agreed with this formulation but designated NH_2Pt as *platiniaque* and N_2H_5Pt as *platinamine.* Otherwise, Gerhardt's views found few adherents.

Jules Reiset (1818–1896)[4] was the first to suggest that the hydrogen atoms of an ammonium group can be replaced not only by metals but also by other ammonium groups, i.e., hydrogen can be substituted partially by metal and partially by ammonium. According to his formulation of the decomposition of his second chloride (*trans*-$[Pt(NH_3)_2Cl_2]$): $3(NH^3PtCl) = 2(ClH, NH^3) + ClH + NPt^3$, he considered that NPt^3 probably existed, if only ephemerally. He did not agree with Berzelius' view (see Introduction to Paper Four) of platinum-ammines as salts of conjugated ammonia. Rather he considered it simpler to assume his chlorides to be true compounds, produced by the most intimate union of ammonia with the metal oxide and possessing the properties of all other bases.

August Wilhelm von Hofmann (1818–1892)[5] specifically applied Reiset's views to the complexes of platinum and cobalt, but in his contribution on the volatile organic bases[8] he frankly confessed that, in his enthusiasm for the ammonium theory, his formulations went somewhat beyond the experimental facts. He represented luteocobalt chloride, $[Co(NH_3)_6]Cl_3$, as

$$Co\left(\begin{array}{l} NH_2—NH_4 \\ | \\ Cl \end{array}\right)_3,$$

and such formulations enjoyed widespread recognition and approval until 1886, when Jørgensen[9] discovered coordination compounds not explicable according to this approach, viz., those formed by the salts of silver, copper, and platinum with pyridine, a tertiary amine that does not contain replaceable hydrogen atoms. Such formulations also did not explain why removal of one of the ammonia molecules completely changes the function of one of the chlorine atoms, rendering it nonionizable.

---

[8] A. W. von Hofmann, *Ann.*, *79*, 11–39 (1851).
[9] S. M. Jørgensen, *J. prakt. Chem.*, [2] *33*, 489–538 (1886).

THOMAS GRAHAM,[10] born in Glasgow, Scotland, on December 21, 1805, entered the University of Glasgow at the age of fourteen in order to study for the ministry, but the lectures of Thomas Thomson convinced him that his calling lay in the field of chemistry. After receiving his Master of Arts degree in 1826, he worked for nearly two years in the laboratory of Thomas Charles Hope at the University of Edinburgh. He then returned to Glasgow where he taught mathematics and chemistry in a private laboratory. In 1829 he became assistant at the Mechanics' Institution and in 1830 was appointed to succeed Alexander Ure as Professor of Chemistry at Anderson's College, later the Royal College of Science and Technology. Graham participated in the organization of the London Chemical Society and in 1841 became its first president. In 1837 he was appointed to succeed Edward Turner as Professor of Chemistry at the University College, London, a post from which he resigned in 1854 to succeed Sir John Herschel as Master of the Mint. This position, which had been held at one time by Sir Isaac Newton, ceased to exist on Graham's death in London on September 16, 1869. Upon the death of John Dalton in 1844, Graham had become the acknowledged dean of English chemists.

Most of Graham's work lay in the field of inorganic and physical chemistry, and he is recognized as the real founder of colloid chemistry. His first original paper *On the Absorption of Gases by Liquids*,[11] which appeared during his twenty-first year, dealt with spontaneous gas movement, a subject which occupied him throughout his career. In fact, almost all his research is but a development, in different directions, of his early works on gaseous diffusion and water of hydration. The famous law of diffusion which bears his name (1829)[12,13] states, as every beginning

---

[10] A. W. Williamson, *J. Chem. Soc.*, *23*, 293–299 (1870); W. Odling, *Report, Board of Regents, Smithsonian Inst.*, 177–216 (1871); reprinted in *Great Chemists*, ed. by E. Farber, Interscience Publishers, New York, 1961, pp. 553–571; T. E. Thorpe, *Essays in Historical Chemistry*, 2nd ed., Macmillan, London, 1902, pp. 206–293; J. C. Irvine, *J. Chem. Educ.*, *7*, 2808–2828 (1930); R. A. Gortner, *J. Chem. Educ.*, *11*, 279–283 (1934); M. Tanaka, *Kagaku* (Kyoto), *14*, 304–307 (1959); J. R. Partington, *A History of Chemistry*, St. Martin's Press, New York, 1964, Vol. 4, pp. 265–270, 272–275, 729–732; M. Speter, in *Das Buch der grossen Chemiker*, ed. by G. Bugge, Verlag Chemie, Weinheim/Bergstrasse, 1965, Vol. 2, pp. 69–77; E. Frame, *Phil. J.*, *7*, 116–127 (1970); G. B. Kauffman, *Dictionary of Scientific Biography*, ed. by C. C. Gillispie, Charles Scribner's Sons, New York, 1972, Vol. 5, pp. 492–495.

[11] T. Graham, *Annals of Philosophy*, *12*, 69 (1826).

[12] T. Graham, *Quart. J. Sci. Arts* (*Royal Inst.*), *27*, 74 (1829); *Phil. Mag.*, *2*, 175, 269, 351 (1833).

[13] A. Ruckstuhl, *J. Chem. Educ.*, *28*, 594–596 (1951); E. A. Mason and B. Kronstadt, *ibid.*, *44*, 740–744 (1967); E. A. Mason and R. B. Evans, *ibid.*, *46*, 359–364, 423–427 (1969); E. A. Mason, *Phil. J.*, *7*, 99–115 (1970).

student of chemistry knows, that the rates of diffusion of gases vary inversely as the square roots of their densities.

Graham's paper *Researches on the Arseniates, Phosphates, and Modifications of Phosphoric Acid* (1833),[14] truly one of the classics of chemistry, established the existence of two new and unexpected classes of substances—polybasic acids and their salts and anhydro acids and their salts. Of this work the eminent German chemist and historian of chemistry Albert Ladenburg has said, "so much has seldom been accomplished by a single investigation." Graham also carried out investigations of phosphorus and phosphine, liquid diffusion, osmosis, the colloidal state, dialysis, and occlusion of gases by metals. His work, usually quantitative, was original in conception, simple in execution, and brilliant in the results to which it led. His original and admirable textbook *Elements of Chemistry* was widely used not only in England but on the continent in its much enlarged multivolume German translation by F. Julius Otto. The following selection is excerpted from the American edition of 1843.—G.B.K.]

[14] T. Graham, *Phil. Trans. Roy. Soc.*, *123*, 253–284 (1833), reprinted as Alembic Club Reprint No. 10, E. and S. Livingstone, Edinburgh, 1961.

3

Thomas Graham

Compounds of Ammonia and Metallic Salts, Supposed to Resemble the Ammonium Compounds

[From *Elements of Chemistry, Including the Applications of the Science in the Arts*, Lea and Blanchard, Philadelphia, 1843, pp. 296–297.]

The whole or a portion of the ammonia absorbed by certain anhydrous salts is retained with great force, and cannot be separated from them by heat. Anhydrous chloride of copper,[1] for instance, absorbs a single equivalent of ammonia with the greatest avidity, and forms a green fusible matter, which the close analogy between copper and hydrogen would lead us to view as analogous in constitution to the compound formed by chloride of hydrogen and ammonia, or chloride of ammonium.[2] It will, therefore, be represented as composed of chlorine united with ammonium, containing an atom of copper, in place of the fourth atom of hydrogen, or as NH_3Cu,Cl,[3] which may be called chloride of cuprammonium. The sulphate of copper, in like manner, retains half an equivalent of ammonia with great force, and forms a compound which may be represented as sulphate of copper combined with sulphate of cuprammonium:

$$CuO,SO_3 + (NH_3CuO + SO_3),$$[4]

[1] [Copper(II) chloride; in "pre-Karlsruhe" notation, CuCl; in modern notation, $CuCl_2$.—G.B.K.]

[2] [Ammonium chloride; in modern notation, NH_4Cl.—G.B.K.]

[3] [In modern notation, $Cu(NH_3)_2Cl_2$.—G.B.K.]

[4] [In modern notation, $Cu(NH_3)_2SO_4$. These formulas are given in terms of the Berzelius dualistic electrochemical theory (1811). See Introduction to Paper Four.—G.B.K.]

which is analogous to the double sulphate of copper and ammonium:

$$CuO,SO_3 + (NH_4,O + SO_3).$$ [5]

Chloride of mercury[6] forms a similar compound with half an equivalent of ammonia:

$$HgCl + NH_3Hg,Cl$$ [7]

analogous to sal-alembroth,[8] or the compound of chloride of mercury and sal-ammoniac:[2]

$$HgCl + NH_4,Cl.$$

A different view of these and the other ammonium compounds is advocated by Dr. Kane.[9] Sal-ammoniac[2] is considered by him as a species of double salt, as amide of hydrogen with chloride of hydrogen, HAd + HCl;[10] and the salt I have named, chloride of cuprammonium, as a corresponding amide of hydrogen with chloride of copper, HAd + CuCl.[3] We agree in supposing these two salts to have the same constitution, but differ as to what that constitution is. To adapt the same explanation to the oxygen acid compounds, such as sulphate of ammonium, Dr. Kane assimilates them to the salts of the magnesian class, which contain two equivalents of oxide. Adopting the hypothesis that two atoms[11] of that class (in which water is included) are equivalent in combination to one of the potash class, he views sulphate of copper, possessing what I term its atom[11] of constitutional water, as a compound of sulphuric acid with a base, which consists of an atom[11] of oxide of copper and an atom[11] of water, and represents it thus:

Sulphate of copper· ·$CuO, HO + SO_3$;[12]

---

[5] [In modern notation, $CuSO_4 \cdot (NH_4)_2SO_4$.—G.B.K.]

[6] [Mercury(II) chloride, in "pre-Karlsruhe" notation, HgCl; in modern notation, $HgCl_2$.—G.B.K.]

[7] [In modern notation, $Hg(NH_3)_2Cl_2$.—G.B.K.]

[8] [A white powder, insoluble in water, formerly used as an antiparasitic; mercury(II) amidochloride, $HgNH_2Cl$.—G.B.K.]

[9] [Robert John Kane (Dublin, Ireland, Sept. 24, 1809—Dublin, Feb. 16, 1890).—G.B.K.]

[10] [For Kane's views of ammonia, its simple salts, and its addition compounds and complexes with metal salts, see his *Elements of Chemistry, Theoretical and Practical, Including the Most Recent Discoveries and Applications of the Science to Medicine, Pharmacy, and to the Arts*, American ed. by J. W. Draper, Harper and Brothers, New York, 1842, Chap. 18. Kane called the radical NH_2 *amidogene*, with the symbol Ad (Kane, *op. cit.*, p. 500).—G.B.K.]

[11] [In modern terminology, molecules.—G.B.K.]

[12] [In modern notation, $CuSO_4 \cdot H_2O$.—G.B.K.]

to which he assimilates the

Sulphate of ammonium · AdH, HO + SO_3.[13]

The hypothesis of the equivalency of two atoms of the magnesian and one of the potash class, has received new support from Dr. Kane's researches; but it is still (in my opinion) too doubtful to form a safe basis for any theoretical superstructure. At the same time, this hypothesis has enabled Dr. Kane to develop many new and interesting relations among the ammoniacal compounds, and may, perhaps, present a closer and more distinct view of the intimate constitution of these bodies, than the ammonium theory exhibits. At present, however, our theories of the constitution of compounds are too uncertain to be regarded otherwise than as artificial aids to facilitate our conception of the manner in which the formation of these bodies occurs, and of the transformations which they undergo; and a theory of constitution is, therefore, adopted more for its convenience than its truth. This state of things leads to the retention of the ammonium theory, which has introduced a degree of simplicity into our views of that particular class of ammoniacal compounds to which it is applicable, that could not easily be exceeded. But its adoption must not be allowed to preclude the consideration of other theories, such as that of Dr. Kane, which facilitate investigations in the meantime, and may prove to be truer to nature in the end.

---

[13] [In modern notation, $(NH_4)_2SO_4$.—G.B.K.]

Carl Ernst Claus (1796–1864)
[From *Izvestiia Instituta po Izucheniiu Platiny i Drugikh Blagorodnykh Metallov*, *6*, 1 (1928).]

INTRODUCTION TO PAPER FOUR

[Shortly after Graham proposed his ammonium theory (1837), the great Swedish chemist Jöns Jacob Berzelius (1779–1848) came forth with his own attempt to explain the constitution of coordination compounds—his so-called conjugate theory. Berzelius, the supreme authority and arbiter of chemical matters during the first half of the nineteenth century, dominated this period with his dualistic electrochemical theory, which he first enunciated in 1811.[1] According to this theory, elements are either electropositive or electronegative, and chemical combination results from the mutual neutralization of opposite charges. However, the compounds thus formed are not necessarily neutral because the opposite charges of the combining atoms are not necessarily equal in magnitude. For example, copper and oxygen combine to form copper oxide, and sulfur and oxygen form sulfur trioxide:

$$\overset{+}{\mathrm{Cu}} + \overset{-}{\mathrm{O}} \longrightarrow \mathrm{CuO}\ (\text{Slightly } +) \quad \text{Primary Compound}$$

$$\overset{+}{\mathrm{S}} + 3\overset{-}{\mathrm{O}} \longrightarrow \mathrm{SO_3}\ (\text{Slightly } -) \quad \text{Primary Compound.}$$

Furthermore, since copper oxide and sulfur trioxide are not neutral, but still carry opposite charges, they can combine with each other to form copper sulfate:

$$\overset{+}{\mathrm{CuO}} + \overset{-}{\mathrm{SO_3}} \longrightarrow \mathrm{CuO \cdot SO_3},\ \text{i.e., } \mathrm{CuSO_4}\ (\text{Slightly } +) \quad \text{Secondary Compound.}$$

Even this salt was not necessarily neutral, and it could combine further to form hydrates, which Berzelius classified as "compounds of higher order," a term later used by Blomstrand and Werner to describe coordination compounds:

$$\mathrm{CuO\overset{+}{\cdot}SO_3} + 5\mathrm{H_2\overset{-}{O}} \longrightarrow \mathrm{CuO \cdot SO_3 \cdot 5H_2O},\ \text{i.e., } \mathrm{CuSO_4 \cdot 5H_2O} \quad \text{Tertiary Compound.}$$

In 1841 Berzelius[2] proposed his conjugate theory, using terms and ideas (*corps copulés*) that he borrowed from the French chemist Charles Gerhardt.[3] According to this theory, he viewed metal-ammines as

---

[1] J. J. Berzelius, *Kongl. Svenska Vetenskaps-Akad. Handl.*, *32*, 169 (1811); *33*, 28, 166, 223 (1812); *J. Phys.*, *73*, 253 (1811); *Ann. Phys.*, *13*, 37 (1812); *Phil. Mag.*, *43*, 3 (1813).

[2] J. J. Berzelius, *Jahresbericht*, *21*, 108 (1841).

[3] C. Gerhardt, *Ann. chim. phys.*, [2] *72*, 184 (1839).

conjugated or copulated compounds consisting of ammonia and a conjugate or copula. The latter cannot be removed by reaction with an acid and neither increases nor decreases the saturation capacity of a base. In other words, a metal in conjugation with ammonia is still capable of combining with other substances. For example, the formula which Berzelius assigned to Reiset's First Chloride ($[Pt(NH_3)_4]Cl_2$) was:

$$\mathrm{Pt}\overline{\mathrm{NH}}^{2}\cdot\overline{\mathrm{NH}}^{4}\cdot\mathrm{Cl},$$

i.e., $Pt(NH_2 \cdot NH_4 \cdot Cl)_2$ or $PtN_2H_4 \cdot N_2H_8Cl_2$. He considered it to be a conjugated compound of platinamide, $PtNH_2$, with the conjugate $N_2H_8Cl_2$ consisting of two molecules of ammonium chloride. Berzelius used barred atoms to represent two atoms or doubled atoms.

Berzelius had introduced his theory of copulae in order to reconcile his electrochemical theory with the new phenomena obtained in the organic field which seemed incompatible with it. In short, the great liberal of the early 1800's had become a conservative and even reactionary force during his later years. As time passed, the number of supporters of his theory diminished and of its opponents increased. More and more complicated and improbable formulas became necessary to make the theory jibe with experimental facts. For example, the French chemist August Laurent ridiculed Berzelius' device in 1855 as follows: "What then is a copula? A copula is an imaginary body, the presence of which disguises all the chemical properties of the compounds with which it is united."[4] In short, Berzelius' stratagem of regarding some compounds as conjugated compounds bore little resemblance to reality in many cases, and although Claus in 1854 and Blomstrand in 1869 in his chain formulas attempted to revive and modify Berzelius' ideas, Berzelius' theory otherwise was of little value.

The next major theory of metal-ammines was proposed by Carl Ernst Claus (1796–1864), Professor of Chemistry at the University of Kazan and later Professor of Chemistry at the University of Dorpat. In 1854 Claus[5] rejected the ammonium theory and suggested a return to Berzelius' view of complexes as conjugated compounds. According to him, the platinum-ammines should be compared not with ammonium salts or with ammonium hydroxide as advocated by adherents of the ammonium theory but with metal oxides. He designated the coordinated ammonia molecule as

[4] A. Laurent, *Chemical Method, Notation, Classification and Nomenclature*, trans. by W. Odling, Cavendish Society, London, 1855, p. 204.

[5] C. Claus, *Beiträge zur Chemie der Platinmetalle*, Festschrift zur Jubelfeier des 50 jährigen Bestehens der Universität Kasan, Glaeser, Dorpat, 1854. Copies of this book, which furnished the foundation for all later work on the platinum metals, later became all but unobtainable. A reprint edition was published by the Chemische Fabrik Braunschweig G.M.B.H. in 1926.

"passive, in contrast to the active, alkaline state in the ammonium salts, where it can easily be detected and replaced by other bases."

Claus' propositions were summarized as three statements in his more widely read paper of 1856, *Ueber die Ammoniummoleküle der Metalle* (Paper Four)[6]:

> (1) If several equivalents of ammonia (from two to six) combine with an equivalent of certain metal chlorides, neutral substances are formed, in which the basic property of ammonia has been destroyed and simultaneously the ammonia can be neither detected by the usual methods nor eliminated by double decomposition.
>
> (2) If the chlorine in these compounds is replaced by oxygen, strong bases are obtained, whose saturation capacity is always determined by the oxygen equivalents contained in them but not by the number of equivalents of ammonia present in them.
>
> (3) The number of equivalents of ammonia entering into these substances is not a random one; as is evident from a number of facts, it is determined by the number of equivalents of water contained in the hydrates of the metal oxides which can enter into such compounds along with the ammonia.

Claus' first postulate was vigorously attacked in 1856 by Carl Weltzien (1813–1870), famed as the organizer of the Karlsruhe Congress,[7] who considered the term "passive" to be indefinite and confusing. Weltzien, one of the leading proponents of the rival ammonium theory, insisted that every part of a molecule affects every other part and therefore no part can be considered passive.[8]

In 1862 Hugo Schiff (1834–1915), Professor of Chemistry at the University of Florence and discoverer of the so-called Schiff bases, attacked not only Claus' first postulate but also his second. With respect to the second postulate, Schiff[9] pointed out that the oxides of the metal-ammines were much stronger bases than the metal oxides themselves. This criticism seems to imply that Schiff was confusing the strength of a base, i.e., the hydroxide ion concentration of its solution, with its saturation capacity, i.e., the number of equivalents of acid with which it can combine.

Claus' third postulate, that is, the close parallel between metal hydrates and metal-ammines, was attacked on the grounds that many hydrates were known for which corresponding ammines were unknown. All of Claus' three postulates reappeared in modified form almost four decades later in Alfred Werner's coordination theory. In fact, Claus'

---

[6] C. Claus, *Ann. Chem.*, *98*, 317–333 (1856).
[7] C. DeMilt, *Chymia*, *1*, 153–169 (1948).
[8] C. Weltzien, *Ann.*, *97*, 19–33 (1856).
[9] H. Schiff, *Ann.*, *123*, 1 (1862).

third postulate closely adumbrates Werner's concepts of the *Koordinationszahl* (coordination number) and of the *Übergangsreihe* (transition series) between metal-ammines and metal salt hydrates. How closely Claus anticipated Werner may be judged by comparing the translation of Claus' paper of 1856 (Paper Four)[6] with Werner's classic paper *Beitrag zur Konstitution anorganischer Verbindungen*[10] of 1893.

Carl Ernst Claus,[11] of German origin, was born in the Baltic city of Dorpat (now Tartu in the Estonian SSR) on January 23, 1796 (Jan. 11, O.S.) and is better known by his Russian name of Karl Karlovich Klaus. Orphaned at an early age, he was forced to earn his own living at the age of fourteen. He became an apprentice to a pharmacist in St. Petersburg, went back to Dorpat in 1815 where he passed the pharmacy examination at the university, and returned to St. Petersburg. In 1817 he moved to Saratov where he spent ten years as a pharmacist's assistant and devoted his leisure studying the flora and fauna of the Volga steppes, on which he became an authority. He opened his own pharmacy in Kazan in 1826, and in 1831 he became an assistant in the Chemistry Department at the University of Dorpat where he studied chemistry and received his bachelor's and master's degrees. He then returned to Kazan where he was appointed adjunct in chemistry (1837) at the university. On receiving his doctorate in pharmacy in 1839, he was made *Extraordinarius Professor* of chemistry and then *Ordinarius Professor* in 1844. In 1852 he moved back to Dorpat to occupy a newly created chair of pharmacy at the university. In 1863 the Russian government sent him to western Europe to visit laboratories and platinum refineries. The following year, on returning home from a scientific meeting in St. Petersburg, he caught a chill, fell ill, and died in Dorpat on March 24, 1864 (Mar. 12, O.S.).

Claus is known primarily as the discoverer of ruthenium,[12] of which he first obtained six grams in 1844 by working up two pounds of platinum residues. He also carried out extensive research on iridium, rhodium, and

[10] A. Werner, *Z. anorg. Chem.*, *3*, 267–330 (1893). For a discussion and English translation see G. B. Kauffman, *Classics in Coordination Chemistry, Part I: The Selected Papers of Alfred Werner*, Dover Publications, Inc., New York, 1968, pp. 5–88.

[11] B. N. Menshutkin, *Izvest. Inst. Izucheniiu Platiny i Drugikh Blagorodn. Metal.*, *6*, 1–11 (1928); *J. Chem. Educ.*, *11*, 226–229 (1934); M. E. Weeks, *ibid.*, *9*, 1017–1034 (1932); *idem, Discovery of the Elements*, 7th ed., Journal of Chemical Education, Easton, Pa., 1968, pp. 415–525; D. McDonald, *A History of Platinum*, Johnson, Matthey and Co., London, 1960, pp. 175–180; *idem, Platinum Metals Review*, *8*, 67–69 (1964); G. B. Kauffman, *Dictionary of Scientific Biography*, ed. by C. C. Gillispie, Charles Scribner's Sons, New York, 1971, Vol. 3, pp. 301–302.

[12] C. Claus, *Memoirs of the Imperial University of Kazan, 1844*, Part 3, 15–200; *J. Pharm. Chim.*, *8*, 381–385 (1845); *Ann.*, *56*, 257 (1845); *Phil. Mag.*, *27*, 230 (1945).

osmium, and in 1854 all his papers on the subject were collected and published as *Beiträge zur Chemie der Platinmetalle*,[5] in a jubilee volume in honor of the fiftieth anniversary of the founding of the University of Kazan. His monograph on the history, chemistry, and applications of the platinum metals, left unfinished upon his death, was eventually published posthumously in 1883 as *Fragment einer Monographie des Platins und der Platinmetalle.*[13] Despite his isolation in a frontier university in eastern Russia, Claus achieved a worldwide reputation for his research on the platinum metals. His second best-known contribution to chemistry is his ammonia theory of metal-ammines (Paper Four).[6]—G.B.K.]

[13] Commissionäre der Kaiserlichen Akademie der Wissenschaften, St. Petersburg, 1883. An extremely rare monograph, presumably only 300 copies were printed. In 1946 the only copy still known to be in existence outside the USSR was the copy of James Lewis Howe in the Chemistry Department Library at Washington and Lee University, Lexington, Virginia.

4

Carl Ernst Claus

The Ammonium Molecules of the Metals

[Translation of "Ueber die Ammoniummoleküle der Metalle," *Annalen der Chemie und Pharmacie*, *98*, 317–333 (1856).]

Let the other side also be heard! [*Audiatur et altera pars!*]

Herr Professor Weltzien[1] has in these *Annalen*[2] rejected my view[3] of, and related formulas for, the rational composition of ammonia-containing platinum bases[4] and has replaced them by other formulas which correspond to the view of these substances which has prevailed until now. Weltzien affirms the ammonium theory[5] for these bases; since I prefer a different way of looking at

[1] [Karl Weltzien (St. Petersburg, Russia, Feb. 8, 1813—Karlsruhe (Baden), Germany, Nov. 14, 1870), Professor of Chemistry at the Karlsruhe Technische Hochschule, is probably best known today for his part in organizing the first international chemical congress, held Sept. 3–5, 1860, at Karlsruhe for the purpose of deciding upon a uniform system of atomic weights.—G.B.K.]

[2] [C. Weltzien, "Ueber die Ammoniummoleküle der Metalle," *Ann.*, *97*, 19–33 (1856).—G.B.K.]

[3] [C. Claus, *Beiträge zur Chemie der Platinmetalle*, Festschrift zur Jubelfeier des 50 jährigen Bestehens der Universität Kasan, Dorpat, 1854, pp. 75, 80. See footnote 5 of the Introduction to this paper.—G.B.K.]

[4] [In modern terminology, platinum-ammine cations.—G.B.K.]

[5] [A theory in which metal-ammines are considered as substituted ammonium compounds (Paper Three). First proposed by Thomas Graham in his book *Elements of Chemistry* (London, 1837), it met with a fair degree of success and was generally accepted by various chemists for several decades, largely because of the modifications of it that were proposed by other chemists such as Gerhardt (*Jahresbericht über die Fortschritte der Chemie*, *3*, 335 (1850)), Wurtz (*Ann. chim. phys.*, [3] *30*, 488 (1850)), Reiset (*Ann. chim. phys.*, [3] *11*, 417 (1844)), Hofmann (*Ann.*, *78*, 253 (1851)), Bödecker (*Ann.*, *123*, 56 (1862)), and Weltzien. It was eventually displaced by two popular ammonia theories—the "molecular compound" concept of Kekulé (see G. B. Kauffman, *Educ. Chem.*, *4*, 11–18 (1967); *J. Chem. Educ.*, *49*, 813–817 (1972)) and the Blomstrand-Jørgensen chain theory (see G. B. Kauffman, *Chymia*, *6*, 180–204 (1960); *J. Chem. Educ.*, *36*, 521–527 (1959); *Ann. Sci.*, *32*, 13–37 (1975)).—G.B.K.]

things, I deny it. May I now be permitted to place the two differing views side by side here so that the reader can decide which of the two has the greater probability.

Weltzien has rejected my view for the following reasons: first, since in his opinion it explains *nothing*, and second, because Wurtz[6] and Hofmann[7] have applied their theory to the platinum bases.

To the first statement I reply: if my view, expressed in formulas, can be brought into agreement with the genesis and the metamorphoses of the platinum bases; if it can be applied unconstrainedly to all known compounds of the type; then it explains more than the ammonium theory, which is found wanting when applied to these compounds. Against the second statement may be used the fact that even though Wurtz and Hofmann are the most competent spokesmen in this matter; even though the introduction of ether radicals[8] into these compounds entitled them to their hypothesis, still, facts have recently come up for discussion which, if they had been known earlier, would certainly have induced these excellent chemists to hesitate in applying their hypothesis to the platinum bases. The introduction of ether radicals[8] into these bases in no way contradicts my view, for if ordinary ammonia can combine with metal oxides with loss of its basicity, why should not an ammonium, in which hydrogen equivalents have been replaced by ether radicals[8] or other radicals, likewise be able to do so?[9]

---

[6] [Charles-Adolphe Wurtz (Strasbourg (Alsace), France, Nov. 26, 1817—Paris, France, May 12, 1884).—G.B.K.]

[7] [August Wilhelm von Hofmann (Giessen, Germany, April 8, 1818—Berlin, Germany, May 2, 1892).—G.B.K.]

[8] [According to Maurice P. Crosland (*Historical Studies in the Language of Chemistry*, Harvard Univ. Press, Cambridge, Mass., 1962, p. 293), "probably the vaguest term in general use in organic chemistry in the first half of the nineteenth century was 'ether.'" "Ether radicals" is probably best translated in the present paper as simply "organic radicals."—G.B.K.]

[9] Weltzien especially emphasizes the fact that ethylamine and methylamine can enter into the platinum bases in place of the ammonia, and he believes that this circumstance is a trustworthy criterion for constituting the composition of these bases according to the ammonium type. It must be admitted that in many cases this behavior can be decisive; namely, if in aniline and other organic bases we can replace hydrogen by ether radicals just as easily as in ammonia, then one is fully justified in concluding that these substances are constituted according to the ammonia type, that is, they are ammonias with substituted hydrogen equivalents. If, on the other hand, ethylamine and methylamine combine with platinum salts in a manner similar to that of ammonia, that is, if they take the place here of ammonia, this is further proof for the homologous constitution of the amines with ammonia, but in no way is it a proof for the ammonium nature of the entire

On the basis of these two statements Weltzien in an extremely ingenious manner has now developed his formulas for these refractory compounds—formulas which, if everything that they assert could be proven, would induce me immediately to throw my hypothesis overboard. Unfortunately, however, they are only makeshift expedients designed to rescue the application of the ammonium theory to these bases; they are not based on even *one* verified fact; they are nothing more than fortuitously solved arithmetic examples, which have not even led to a result favorable to the author,[10] since several of the postulated formulas do not even agree with the ammonium theory and directly contradict other consistently worked-out formulas. In short, from the entire exposition it is evident that a consistent application of the ammonium theory to the facts under discussion cannot be forced.

As proof for this claim, the author's[10] formulas and their consequences should now be examined more closely.

$$\text{Mercurous Chloride-Ammonia} = N \begin{Bmatrix} H_3 \\ Hg \\ Hg \end{Bmatrix} Cl;^{11}$$

$$\text{Ferric Chloride-Ammonia} = N \begin{Bmatrix} H_3 \\ Fe \\ Fe \end{Bmatrix} Cl_3.^{12}$$

These few examples should be enough to show the manner of formulation. Will anyone again recognize the ammonium compound type in these formulas? These compounds were summarily dismissed since the author[10] has apparently found in them a composition unfavorable to his view. He even expresses this, inasmuch

compound. H. Rose [1795–1864] has prepared the compound $5NH_3$, $PCl_3[5NH_3 \cdot PCl_3]$.[9a] It is highly probable, if not certain, that it is possible to prepare a substance $5C_2H_5N$, $PCl_3[5CH_3NH_2 \cdot PCl_3]$. Who indeed could claim that the latter substance is an ammonium compound!

[9a] [*Ann. Phys. Chem.*, *24*, 295–341 (1832), p. 308; *28*, 529–550 (1833). Also earlier, H. Davy, *Ann. Phys.*, *39*, 3–33 (1811), p. 6, and later, A. Besson, *Compt. rend.*, *111*, 972–974 (1890). J. Persoz reported the compound $PCl_3 \cdot 4NH_3$ (*Ann. chim. phys.*, [2] *44*, 315–325 (1830), p. 321).—G.B.K.]

[10] [Weltzien.—G.B.K.]

[11] [All the formulas in this paper naturally antedate the Karlsruhe Congress. In modern notation, $HgCl \cdot 2NH_3$.—G.B.K.]

[12] [$FeCl_3 \cdot 2NH_3$.—G.B.K.]

as he notes: "Mercurous chloride-ammonia [*Quecksilberchlorür-Ammoniak*] could be regarded as a compound containing one molecule of mercury ammonium [*Quecksilberammonium*] and one equivalent of mercury to one equivalent of chlorine. The compound of ammonia with ferric chloride, then, would be constituted in a similar manner." These formulas are thus intended to indicate a similar constitution!

There now follow the formulas of such compounds as may be regarded as belonging to the ammonium type only upon the assumption of certain hypotheses.

$$\mathrm{CoCl} + 2\mathrm{NH_3}\,{}^{13} = \mathrm{N}\left\{\begin{matrix}\mathrm{H}\\ \mathrm{H}\\ \mathrm{NH_4}\\ \mathrm{Co}\end{matrix}\right\}\mathrm{Cl}; \qquad \mathrm{PtCl} + 2\mathrm{NH_3}\,{}^{14} = \mathrm{N}\left\{\begin{matrix}\mathrm{H}\\ \mathrm{H}\\ \mathrm{NH_4}\\ \mathrm{Pt}\end{matrix}\right\}\mathrm{Cl};$$

$$\mathrm{PdCl} + 2\mathrm{NH_3}\,{}^{15} = \mathrm{N}\left\{\begin{matrix}\mathrm{H}\\ \mathrm{H}\\ \mathrm{NH_4}\\ \mathrm{Pd}\end{matrix}\right\}\mathrm{Cl}.$$

$$\mathrm{PtCl} + \mathrm{NH_3}\,{}^{16} = \mathrm{N}\left\{\begin{matrix}\mathrm{H}\\ \mathrm{H}\\ \mathrm{H}\\ \mathrm{Pt}\end{matrix}\right\}\mathrm{Cl}; \qquad \mathrm{PtO, NH_3} + \mathrm{SO_3}\,{}^{17} = \mathrm{N}\left\{\begin{matrix}\mathrm{H}\\ \mathrm{H}\\ \mathrm{H}\\ \mathrm{Pt}\end{matrix}\right\}\mathrm{O} + \mathrm{SO_3};$$

$$\mathrm{PtCl_2} + \mathrm{NH_3}\,{}^{18} = \mathrm{N}\left\{\begin{matrix}\mathrm{H}\\ \mathrm{H}\\ \mathrm{H}\\ \mathrm{Pt}\end{matrix}\right\}\mathrm{Cl_2}; \qquad \mathrm{PtO_2, NH_3} + 2\mathrm{SO_3}\,{}^{19} = \mathrm{N}\left\{\begin{matrix}\mathrm{H}\\ \mathrm{H}\\ \mathrm{H}\\ \mathrm{Pt}\end{matrix}\right\}\mathrm{O_2} + 2\mathrm{SO_3};$$

$$\text{Gros' Salt } \mathrm{PtCl_2} + 2\mathrm{NH_3}\,{}^{20} = \mathrm{N}\left\{\begin{matrix}\mathrm{H}\\ \mathrm{H}\\ \mathrm{NH_3}\\ \mathrm{Pt}\end{matrix}\right\}2\mathrm{Cl}.$$

---

[13] [$CoCl_2 \cdot 4NH_3$.—G.B.K.]
[14] [$PtCl_2 \cdot 4NH_3$.—G.B.K.]
[15] [$PdCl_2 \cdot 4NH_3$.—G.B.K.]
[16] [$PtCl_2 \cdot 2NH_3$.—G.B.K.]
[17] [$PtSO_4 \cdot 2NH_3$.—G.B.K.]
[18] [$PtCl_4 \cdot 2NH_3$.—G.B.K.]
[19] [$Pt(SO_4)_2 \cdot 2NH_3$.—G.B.K.]
[20] [$PtCl_4 \cdot 4NH_3$; actually *trans*-$[Pt(NH_3)_4Cl_2]Cl_2$.—G.B.K.]

The bases which contain more than two equivalents of ammonia are formulated in a particularly peculiar manner.

Rhodium Base and its Salts

$$Rh_2O_3,\ 5NH_3:\text{[21]}\ \left.\begin{matrix} N\left\{\begin{matrix} H \\ H \\ NH_4 \\ Rh \end{matrix}\right. \\ N\left\{\begin{matrix} H \\ NH_4 \\ NH_4 \\ Rh \end{matrix}\right. \end{matrix}\right\} O_3.$$

Iridium Base

$$Ir_2O_3,\ 5NH_3:\text{[22]}\ \left.\begin{matrix} N\left\{\begin{matrix} H \\ H \\ NH_4 \\ Ir \end{matrix}\right. \\ N\left\{\begin{matrix} H \\ NH_4 \\ NH_4 \\ Ir \end{matrix}\right. \end{matrix}\right\} O_3.$$

Luteocobalt-ammine Salt

$$Co_2O_3,\ 6NH_3 + 3SO_3:\text{[23]}\ \left.\begin{matrix} N\left\{\begin{matrix} H \\ NH_4 \\ NH_4 \\ Co \end{matrix}\right. \\ N\left\{\begin{matrix} H \\ NH_4 \\ NH_4 \\ Co \end{matrix}\right. \end{matrix}\right\} O_3 + 3SO_3 = \left.\begin{matrix} N\left\{\begin{matrix} H \\ H \\ NH_4 \\ Co \end{matrix}\right. \\ N\left\{\begin{matrix} H \\ NH_4 \\ NH_4 \\ Co \end{matrix}\right. \end{matrix}\right\} O_3 + 3SO_3 + N$$

Even this small number of examples should suffice. These formulas, which, in highly complicated form, force the compounds in question into a completely new ammonium theory, cannot be regarded as an expression of their composition unless the following hypotheses are simultaneously assumed to be true.

(1) That there exists an ammonium in which hydrogen can be replaced by equivalents of metal.

(2) That, in addition to the replacement of hydrogen by metal, replacements by ammonium are possible, indeed without limit up to the complete replacement of all hydrogens, so that there can exist ammoniums which in one equivalent can contain two to six and even more equivalents of nitrogen.

(3) That an equivalent of ammonium which contains metal or ammonium can combine with one or two equivalents of oxygen to

[21] [$Rh_2O_3 \cdot 10NH_3$.—G.B.K.]
[22] [$Ir_2O_3 \cdot 10NH_3$.—G.B.K.]
[23] [$Co_2(SO_4)_3 \cdot 12NH_3$; actually $[Co(NH_3)_6]_2(SO_4)_3$.—G.B.K.]

form an ammonium protoxide and deutoxide[24] and can saturate one or two equivalents of acid, respectively.

(4) That two equivalents of an ammonium, in which hydrogen is replaced partly by metal, partly by ammonium, combine with three equivalents of oxygen to form a sesquioxide and can form salts which are similar to the ordinary sesquioxide salts.

(5) That ammonia, by acting upon chlorides, oxides, or oxygen salts of certain metals and also simultaneously acting upon itself, is subject to innumerable metamorphoses to become an ammonium and can then cause substitutions of the most complicated kind.

The final assumption, although not proven by any facts, still has some intrinsic probability, but only within the meaning of the ammonium theory, for only the metal-containing ammonia compounds have given rise to this. Now if another more satisfactory explanation for these compounds is found, that assumption automatically collapses.

All other hypotheses are substantiated neither in the ammonium theory as it has existed until now nor by any actual fact; they are borrowed from a lively imagination in order to formulate the composition of our compounds. If these arbitrary hypotheses could be proven, how the horizons of the ammonium theory would then expand and serve as a magnificent triumph for its admirers! But that there is little hope for this is already proven by those first-cited formulas, which can in no way be integrated. On the contrary, these hypotheses carry the seed of death within themselves since they gravitate in the opposite direction.

On the other hand, if we consider the relationship of the ammonium theory to the organic bases, both those completely formed in nature and those artificially produced, then we find here a complete analogy in composition, in behavior, and in most of the relationships of the salts of these bases with ordinary ammonia salts. Each new discovery only confirms this analogy. Almost every substitution which can be accomplished in ammonia or its salts can be proven in the organic bases, indeed not substitutions made merely

[24] [Thomas Thomson (*System of Chemistry*, 3rd ed., Bell and Bradfute, Edinburgh, 1807, Vol. I) distinguished oxides as follows: "The *protoxide* of a metal will denote the metal combined with a minimum of oxygen or the *first oxide* which the metal is capable of forming; *deutoxide* will denote the second oxide of the metal."—G.B.K.]

on paper but those achieved by infallible experiment. We can arbitrarily limit or expand these substitutions. All salts of organic bases have a homologous series in the ordinary ammonia salts, for:

$$\left.\begin{matrix} C_4H_5 \\ C_4H_5 \\ C_4H_5 \\ C_4H_5 \end{matrix}\right\} N + Cl\text{[25]} = \left.\begin{matrix} C_{12}H_5 \\ H \\ H \\ H \end{matrix}\right\} N + Cl\text{[26]} = \left.\begin{matrix} H \\ H \\ H \\ H \end{matrix}\right\} N + Cl.$$

In short, the ammonium theory finds its complete justification in the behavior and composition of these compounds. However, we now find among the great number of carefully investigated organic bases not a single example in which it can be proven that the assumption of an ammonium in which H is replaced by NH_4, much less of an ammonium with two to three equivalents of oxygen as ammonium oxide can be justified. Has anyone prepared an ordinary ammonia salt of this type? As long as ammonium equivalents cannot be introduced into ammonia or its salts, the assumption of such substitutions should be rejected. According to the ammonium theory, how easily should an ammonium chloride which contains two equivalents of hydrogen be formed if platinous chloride-ammonia and ammonium chloride had the same constitution; for just as easily as the compound $\left.\begin{matrix} NH_4 \\ H_2 \\ Pt \end{matrix}\right\} N + Cl$[27] can be formed from $\left.\begin{matrix} H_3 \\ Pt \end{matrix}\right\} N + Cl$[28] and NH_3, so should the compound $\left.\begin{matrix} NH_4 \\ H_3 \end{matrix}\right\} N + Cl$ also be formed from $\left.\begin{matrix} H_3 \\ H \end{matrix}\right\} N + Cl$ and NH_3.

Indeed, in this way, by basing further conclusions on this uncertain assumption, an ammonium chloride should be formed which should contain three equivalents of chlorine and ten equivalents of nitrogen in one equivalent; for the following formula can be written very easily on paper:

---

[25] $[[(C_2H_5)_4N]Cl$.—G.B.K.]

[26] $[[(C_6H_5)NH_3]Cl$.—G.B.K.]

[27] $[Pt(NH_4)NH_2Cl_2$.—G.B.K.]

[28] $[PtNH_3Cl_2$.—G.B.K.]

$$\left.\begin{array}{l}\left.\begin{array}{l}\mathrm{NH_4}\\ \mathrm{NH_4}\\ \mathrm{NH_4}\\ \mathrm{NH_4}\end{array}\right\}\mathrm{N}\\ \left.\begin{array}{l}\mathrm{NH_4}\\ \mathrm{NH_4}\\ \mathrm{NH_4}\\ \mathrm{NH_4}\end{array}\right\}\mathrm{N}\end{array}\right\} + \mathrm{Cl_3} = \left.\begin{array}{l}\left.\begin{array}{r}\mathrm{NH_4}\\ \mathrm{H}\\ \mathrm{H}\\ \mathrm{Rh}\end{array}\right\}\mathrm{N}\\ \left.\begin{array}{r}\mathrm{NH_4}\\ \mathrm{NH_4}\\ \mathrm{H}\\ \mathrm{Rh}\end{array}\right\}\mathrm{N}\end{array}\right\} + \mathrm{Cl_3} = \text{Chlorine Compound of the Rhodium Base}$$

In the series of ordinary ammonia salts, this ammonium chloride would be the analogue of the chlorine compound of the rhodium base and of the iridium base and also of the chloride of Fremy's[29] roseocobaltammine;[30] for according to this method of formulation, it is immaterial whether one, two, or all equivalents of the hydrogen in the ammonium are substituted by ammonium. Other chemists besides me, who have occupied themselves in a *practical* way with this matter, have also understood that the ammonium theory does not suffice for the compounds in question. Fremy has not relied on this theory for his peculiar ammonia-containing cobalt salts, but by the designation "cobalt-ammines" [*Kobaltiake*] he has indicated a method of viewing their composition which coincides with mine. After having discussed his experiments on a new platinum base which is the decomposition product of pyridine hydrochloride with platinum chloride, Anderson[31,32] closes with the remark: "The views of these bases held until now must be altered somewhat." Also Otto,[33]

---

[29] [Edmond Fremy (Versailles, France, Feb. 29, 1814—Paris, France, Feb. 3, 1894), Professor of Chemistry at the École Polytechnique, carried out some of the earliest experimental research on cobalt-ammines and originated the color system of nomenclature for coordination compounds (luteo, purpureo, xantho, roseo, croceo, praseo, etc.) (*Ann. chim. phys.*, [3] *35*, 257 (1852); *J. prakt. Chem.*, *57*, 95 (1852)).—G.B.K.]

[30] [$[Co(NH_3)_5H_2O]Cl_3$.—G.B.K.]

[31] [Thomas Anderson (Leith, Scotland, July 2, 1819—Chiswick, England, Nov. 2, 1874), Professor of Chemistry at the University of Glasgow, investigated the nitrogenous bases in bone oil and discovered pyridine, lutidine, picoline, and collidine.—G.B.K.]

[32] *Ann.*, *96*, 205 [199–205 (1855)]. If the double salt of pyridine hydrochloride with platinum chloride = $C_{10}H_6NCl + PtCl_2$ [in modern formulation, $(C_5H_5NH)_2[PtCl_6]$] is boiled for a long time with water, it is converted, with escape of HCl into the chlorine compound of a new platinum base, $C_{10}H_5NPt + Cl_2$ [in modern formulation, *cis*-$[Pt(C_5H_5N)_2Cl_4]$]. Again a fact which supports my view.

[33] [Friedrich Julius Otto (1809–1870).—G.B.K.]

in his excellent textbook,[34] second edition, points out a number of disadvantages in the use of that theory for the platinum bases, and in the third edition he definitely concurs with my view. Finally, if we consider what Hofmann,[7,35] the founder of the formulas for the platinum bases within the meaning of the ammonium theory and of the substitutions with NH_4, says about this, then we become convinced that he too did not value these formulas very highly, for while discussing Reiset's[36] First Base,[37] he expresses himself as follows: "At present we have no criterion for a definite view on the way in which this ammonia equivalent is bound to the platosammonium oxide.[38] If one applies to these substances Graham's[5,39] remark that the ammonia, when bonded directly to hydrogen-containing substances, can be regarded as ammonium which replaces one equivalent of hydrogen, then one could regard the platosamine as an ammoniopIatosammonium oxide, composed according to the formula $\left.\begin{matrix}H_2\\ NH_4\\ Pt\end{matrix}\right\}NO$. *However, it cannot be denied that this viewpoint requires further experimental confirmation.*" Graham was completely correct in this remark if he had HS,[40] HCl, HCy[41] in mind, but whether this remark could be applied to an already complete ammonium is very much to be doubted. Furthermore, Hofmann says in another place, which Weltzien[42] also cites: "He willingly admits that here theory runs somewhat ahead of experiment and that in his admiration for ammonium he has gone too far."

[34] [Thomas Graham and Friedrich Julius Otto, *Ausführliches Lehrbuch der Chemie*, Vol. I, *Lehrbuch der physikalischen und theoretischen Chemie*, Friedrich Vieweg und Sohn, Braunschweig, 1855.—G.B.K.]

[35] *Handwörterbuch der reinen und angewandten Chemie von Liebig und Poggendorf* [*f*], Supplementband, 1850, p. 473.

[36] [Jules Reiset (1818–1896).—G.B.K.]

[37] [$[Pt(NH_3)_4](OH)_2$ (J. Reiset, *Compt. rend.*, *10*, 870–872 (1840); *11*, 711–712 (1840); *18*, 1100–1105 (1844); *Ann. chim. phys.*, [3] *11*, 417–433 (1844)).—G.B.K.]

[38] [Platosammine chloride is *trans*-$[Pt(NH_3)_2Cl_2]$.—G.B.K.]

[39] [Thomas Graham (Glasgow, Scotland, Dec. 21, 1805—London, England, Sept. 16, 1869), Professor of Chemistry at Anderson's College and later University College, London, and first president of the Chemical Society, originated the ammonium theory of metal-ammines (Paper Three).—G.B.K.]

[40] [H_2S.—G.B.K.]

[41] [HCN.—G.B.K.]

[42] [A. W. Hofmann, "Beiträge zur Kenntnis der flüchtigen organischen Basen"] *Ann.*, *79*, 11[–39 (1851)].

Now the facts cited up to now and the newer facts are sufficient for us to conclude that another view of the composition of these substances is justified, provided that this view eliminates the objections to the earlier one, that it can be brought into agreement with all heretofore known facts, and that it can be extended to all compounds belonging here without doing violence to science. In my *Contributions to the Chemistry of the Platinum Metals*,[3] I have attempted to pave the way for such a view through the formulas advanced there, although I did not articulate its entire content. I was especially concerned with emphasizing the experimental findings; the consequences then follow automatically. These findings can be summarized in the following three statements: (1) If several (from two to six) equivalents of ammonia combine with one equivalent of certain metal chlorides, neutral substances are formed, in which the basic property of the ammonia is destroyed, and simultaneously the ammonia can be neither detected by ordinary means nor eliminated by double decomposition. From this it now follows that the ammonia here is in a different state than in the ordinary ammonia salts. I have named this state passive, in contrast to the active, alkaline state in the ammonia salts, where it can easily be detected and replaced by other bases. This nomenclature is a nonessential one, with which I link no further theoretical view and on whose retention I in no way insist for science.[43]—(2) If, in these compounds, chlorine is replaced by oxygen, strong oxygen bases are obtained, whose saturation capacity is always determined by the oxygen equivalents contained in them but not by the number of equivalents of ammonia present in them.—(3) The number of ammonia equivalents entering into these substances is not an accidental one;[44] as is evident from a number of facts, it is determined by the number of water equivalents contained in the hydrates of the metal oxides which can enter into such compounds along with the ammonia.[45] This statement, important as it is, cannot yet be assumed to be generally valid because of lack of the requisite number of facts.

---

[43] Weltzien has also proclaimed a death sentence on this harmless expression.

[44] [This statement anticipates to some extent Werner's concept of the coordination number.—G.B.K.]

[45] [Werner later showed the close relationship between metal-ammines and metal hydrates and demonstrated the existence of transition series (*Übergangsreihen*) linking the two types of compounds.—G.B.K.]

Let us now discuss the conclusions that can be drawn from these statements. In these compounds, therefore, ammonia is bound in a manner different from that in the ammonia salts; on this point all factions agree. The new bases with five equivalents and especially those with three equivalents of ammonia occurring in them have proven that it cannot be in the state of an ammonium salt with substituted hydrogen equivalents. Therefore these bases too cannot be compared with ammonium oxide, nor can their salts be compared with the ammonium oxide salts. But if we compare them with the metal oxides, such a striking analogy emerges that not a single fact contradicting this analogy can be discovered, for these bases chemically behave exactly like the very metal oxides which are contained in them. If the metal which occurs in these substances possesses two oxidation states, then we can also prepare two different ammonia-containing bases corresponding to these oxidation states, bases which behave just like the oxides themselves in that they saturate one or two equivalents of acid, depending on their oxygen content. If the metal preferentially forms a sesquioxide with three equivalents of oxygen, then bases of this composition are also obtained, and these too, like their oxides, saturate three equivalents of acids. These bases can also form basic salts in addition to the ordinary neutral salts, as can be proven by the salts of the type prepared by Fremy and Gerhardt.[46] In short, these substances behave like metal oxides, and the ammonia is thus in reality *passive*. These important, decisive relationships have been little noted by chemists until now, partly because too few facts were yet available and partly because everyone was too captivated by the attractive ammonium theory. In view of the new facts, however, this beautiful theory has become impossible for the platinum bases. For who indeed could maintain that there are basic ammonium oxide salts or an ammonium sesquioxide and ammonium dioxide?

As already stated, there is probably no longer any doubt that the ammonia in these compounds is present in a manner different from that of an ammonium with substituted hydrogen equivalents. But what kind of manner might this be?

According to the present state of knowledge, it seems to me that no other remedy remains except to assume that the ammonia occurs

---

[46] [Charles Frédéric Gerhardt (Strasbourg (Alsace), France, Aug. 21, 1816—Strasbourg, Aug. 19, 1856).—G.B.K.]

here in a *copulated* compound [*gepaarte Verbindung*][47] and that the radicals of these compounds with ammonia are copulated metals. This assumption contains nothing strange to science; the concept of copulation [*Paarung*] has been introduced into it, and the behavior of the ammonia in these compounds corresponds to this concept. What is new and unfamiliar is the copula [*Paarling*] itself, the ammonia, which here appears as such for the first time. Such copulas with metals do not constitute unusual phenomena; indeed we have a considerable number of similarly constituted metal compounds with ether radicals,[8] such as cacodyl,[48] ethyltin,[49] methylmercury[49] and ethylmercury,[49] etc. Most of these contain more than one equivalent of the copula in a manner similar to that by which the platinum bases contain the ammonia. Can the ammonia be denied the functions of a radical? I do not believe so; for when facts such as ours are available, it can be called a radical with the same justification as the ammonium. That a juxtaposition of the ammonia with the ether radicals[8] involves nothing absurd is proven by the fact that the oxides of those hydrocarbon-containing metals behave exactly like the oxides themselves, as do the ammonia-containing platinum bases. It was not very long ago that Dumas,[50] in formulating his first ether theory, placed ammonia parallel to etherin.[51]

---

[47] [The term "copula" had originally been used by Gerhardt (*Ann. chim. phys.*, [2] *72*, 184 (1839)) to denote the organic part of a molecule that might be joined to an inorganic part. In 1841 Berzelius applied these ideas of "copulated" or "conjugated" compounds to coordination compounds in what was one of the earliest systematic attempts to explain their constitution (see also p. 47).—G.B.K.]

[48] [Tetramethyldiarsine radical, $(CH_3)_4As_2$, whose discovery by Robert Bunsen (*Ann. Phys.*, *40*, 219 (1837); *42*, 145 (1837); *Ann.*, *37*, 1 (1841); *42*, 14 (1842); *46*, 1 (1843)) was considered by Edward Frankland to be "a most remarkable confirmation of the theory of organic radicals, as propounded by Berzelius and Liebig."—G.B.K.]

[49] [The discovery of these "radicals," among the first organometallic compounds, by Edward Frankland led directly to his concept of valency.—G.B.K.]

[50] [Jean Baptiste André Dumas (Alais, France, July 14, 1800—Cannes, France, April 11, 1884), Professor of Chemistry at the École Polytechnique, the École de Médecine, and the Sorbonne, made many contributions to organic chemistry.—G.B.K.]

[51] [Dumas and Boullay (*Ann. chim. phys.*, [2] *37*, 15–23 (1828)) attempted to relate alcohol, ether, ethyl chloride, and other compounds by supposing them to consist of a radical, C_2H_4, called "etherin" by Berzelius (*Ann.*, *3*, 282 (1832)). According to Kekulé, this theory was important in being the first attempt to connect a series of substances by a common bond, but it was never applied to many substances and it obstructed the development of the radical theory.—G.B.K.]

Therefore the hypothesis of the copulation of ammonia equivalents in these bases has an inductive basis equally as justified as the hypothesis of the ammonium molecules. However, it has the advantage over the latter, that without further hypothetical suppositions it explains the genesis and metamorphoses, in short, all phenomena, in a much more satisfactory manner than the latter view. Also, it can be applied unconstrainedly to all compounds belonging here, and it presents no difficulty in the formulation of formulas. Furthermore, it acquaints us with a new, hitherto unnoticed, relationship of the ammonia whose significance could become very important.

The following may serve as proof for the facts cited: The meaningless formulas $N\left\{\begin{matrix}H_3\\Hg\\Hg\end{matrix}\right\}Cl$ [11] and $N\left\{\begin{matrix}H_3\\Fe\\Fe\end{matrix}\right\}Cl_3$ [12] are transformed into $\widehat{NH_3, Hg_2} + Cl$ and $\widehat{NH_3, Fe_2} + Cl_3$; furthermore $N\left\{\begin{matrix}H_3\\Pt\end{matrix}\right\}Cl_2$ and $N\left\{\begin{matrix}H_2\\NH_3\\Pt\end{matrix}\right\}Cl_2$ become $\widehat{NH_3, Pt} + Cl_2$ and $2\widehat{NH_3, Pt} + Cl_2$.

$$\left.\begin{matrix}N\left\{\begin{matrix}H_2\\NH_4\\Rh\end{matrix}\right.\\N\left\{\begin{matrix}H\\NH_4\\NH_4\\Rh\end{matrix}\right.\end{matrix}\right\}Cl_3 \quad \text{becomes } 5\widehat{NH_3, Rh_2} + Cl_3.$$

So much for the formulas; the following concerns the explanation of the phenomena.

Magnus'[52] Salt $\widehat{NH_3, Pt} + Cl$,[53] which is most indifferent toward

[52] [Heinrich Gustav Magnus (Berlin, Germany, May 2, 1802—Berlin, April 4, 1870), Professor of Technology and Physics at the University of Berlin, worked with Berzelius at Stockholm during 1827 and 1828, where he discovered the salt which bears his name and worked on the compound later known as Zeise's Salt. See Paper One.—G.B.K.]

[53] [The first platinum-ammine to be discovered, $[Pt(NH_3)_4][PtCl_4]$, tetraammineplatinum(II) tetrachloroplatinate(II) (G. Magnus, *Ann. Phys.*, *14*, 239–242 (1828); see Paper One of this volume; G. B. Kauffman, *Platinum Metals Rev.*, *20*, 21–24 (1975)).—G.B.K.]

dilute acids and alkalies, dissolves very readily in very dilute chlorine water with a light yellow color. The reason for this phenomenon cannot easily be discovered according to the ammonium theory, whereas according to the other view it can easily be seen that the chlorine converts the difficultly soluble green platinous chloride into the readily soluble yellow platinic chloride, completely independently of the ammonia which is present. The latter salt is converted immediately into the compound $NH_{\widehat{3}, Pt} + Cl_2$, which can be obtained from the solution in light yellow crystals.[54] If Magnus' Salt is treated with an excess of strong chlorine water, then along with the conversion of platinous chloride into platinic chloride and the evolution of nitrogen chloride, the decomposition of the copula, of the ammonia, begins, and the final result of the action is free platinic chloride, hydrochloric acid, and some ammonium chloroplatinate.—If the green salt is treated for a longer time with ammonia, a second equivalent of ammonia enters the compound, and this compound dissolves to form a colorless solution, from which Reiset's Salt, the chlorine compound of his first base $= 2NH_{\widehat{3}, Pt} + Cl$ may be obtained.[55,56] If this salt is treated carefully with chlorine water, it is converted into the so-called chlorine compound of [J.] Gros' base,[57] while the platinous chloride becomes platinic chloride. Upon continued action of chlorine, one equivalent of the copula is decomposed with phenomena similar to the above, and the earlier-mentioned salt $NH_{\widehat{3}, Pt} + Cl_2$[58] is again obtained, and this is finally decomposed into platinic chloride and ammonium chloroplatinate. How simple and logical the explanation of the process is here. But if we try to explain these phenomena within the meaning of the ammonium theory, we encounter self-created difficulties. The ammonium chloride in which one equivalent of hydrogen is replaced by platinum $= \left.\begin{matrix}H_3\\ Pt\end{matrix}\right\}N, Cl$ is converted by the action of chlorine into

---

[54] [According to C. Gerhardt (*Ann.*, *76*, 314, 307–316 (1850)), Magnus' Green Salt is oxidized by chlorine first to *trans*-$[Pt(NH_3)_4Cl_2][PtCl_4]$ and then to *trans*-$[Pt(NH_3)_4Cl_2][PtCl_6]$ (in old notation, $NH_{\widehat{3}, Pt} + Cl_2$).—G.B.K.]

[55] This salt, with ammonium chloride, is found in considerable amounts in the mother liquor which is obtained in the preparation of Magnus' Salt; by certain manipulations, in addition to this salt, one can also obtain Peyrone's white and red compound [*cis*-$[Pt(NH_3)_2Cl_2]$], which is isomeric with Magnus' Salt.

[56] [$[Pt(NH_3)_4][PtCl_4] + 4NH_3 \rightarrow 2\{[Pt(NH_3)_4]Cl_2\}$.—G.B.K.]

[57] [*trans*-$[Pt(NH_3)_4Cl_2]Cl_2$.—G.B.K.]

[58] [*trans*-$[Pt(NH_3)_4Cl_2][PtCl_6]$.—G.B.K.]

$\left.\begin{matrix} H_3 \\ Pt \end{matrix}\right\} N, Cl_2$, an ammonium chloride which contains two equivalents of chlorine. This explanation is not in agreement with the ammonium theory, for in this theory no ammonium chloride of this type has hitherto been formulated. Furthermore, if ammonia reacts with that ammonium chloride with one equivalent of chlorine, then the peculiar case is said to occur in which the ammonia removes hydrogen from an already complete ammonium in order to become ammonium itself and to replace this hydrogen. The inexplicable explanations are repeated in the further course of the proceedings. And this view, in Weltzien's opinion, explains everything, but mine nothing!

The third statement, if it is to prove of general validity in the future, would be of special interest; it would bring us closer to the cause which determines that metal oxides copulate with one to two equivalents of ammonia in one case and with several equivalents in others. This circumstance would depend upon a regular compound type which reveals itself in the compounds of the metal oxides with water and ammonia. In any case, the agreement of the unusually composed hydrous rhodous sesquioxide $= Rh_2O_3 + 5aq.$ with $Rh_2O_3 + 5NH_3$ is very peculiar. Most probably the hydrous iridous sesquioxide Ir_2O_3 also contains five equivalents of water,[59] to which is added the hydrous chromic oxide $Cr_2O_3 + 5aq.$, so that this number of water equivalents does not exist as an isolated fact. From the composition of the platinum bases, in connection with the facts just discussed, the conclusion can now be drawn that the oxides of the platinum metals can form hydrates with variable water content, inasmuch as an equivalent of an oxide can combine with one or two equivalents of ammonia to form bases. The preparation of the hydrates $PtO + HO$, $PtO + 2HO$, $PtO_2 + HO$, and $PtO_2 + 2HO$[60] should therefore be attempted. That one and the

[59] Until now I have not succeeded in preparing pure hydrated iridous sesquioxide; otherwise I would already have determined its composition. For if the lower sesquichloride is treated with a little potash, indeed an oxide of light gray, opalescent greenish color is obtained. But the oxide still contains undecomposed lower chloride, which can neither be washed out nor removed in another way. If more potash is used, then no precipitate at all is obtained since the hydrated lower sesquioxide is even more readily soluble in potash than the hydrated rhodous sesquioxide. However, I hope to be able to determine the water content in an indirect manner. Besides, this oxidation state is just as oxyphoric [susceptible to oxidation] as the hydrous ferrous oxide; it is immediately converted into the blue iridium oxide.

[60] [In these "pre-Karlsruhe" formulas, HO represents water (H_2O).—G.B.K.]

same metal oxide can combine with water in different proportions has been definitely proven during my work with the platinum metals, for in addition to the light yellow hydrous rhodous sesquioxide[61] and the light gray-greenish hydrous iridous sesquioxide[62] there also exist for both metals in the same oxidation state black hydrates,[63] which are obtained by decomposing the respective chlorine compounds with alcoholic potash.[64] These hydrates have not yet been analyzed, but I do not believe that I am wrong if I provisionally assume three equivalents of water in them so that they may be the normal hydrates, in which the oxygen content of the oxide and water of hydration is equal. If the bases with three equivalents of ammonia corresponding to these hydrates should now be successfully prepared, then the third of my empirical statements would also be proven.[65] What a wide field for investigations opens here for the chemist! Whereas if we are satisfied with the ammonium theory and the motley formulas for our compounds, all is over and done with.

Finally I must note that the formulas with the copula notation [*Zeichen der Paarung*] for these compounds were first used by Otto.[33] I mention this here so that the reader will not believe that I am appropriating the credit for this. In spite of this, I believe that I have the right to make use of this manner of expression and these formulas, since they agreed with my first communication on this matter and were so understood by Otto also. Furthermore, I herewith declare that it was in no way my intention to offend Herr Weltzien, but that I merely wished to defend the view of which I have become convinced.

---

[61] [See p. 41 of reference 3.—G.B.K.]

[62] [C. Claus, *J. prakt. Chem.*, *80*, 301 (1860); *Bull. de la Classe physico-mathématique de l'Académie Impériale des Sciences de St.-Pétersbourg*, [3] *2*, 175 (1860).—G.B.K.]

[63] [C. Claus, *J. prakt. Chem.*, *80*, 296 (1860); *Bull. de la Classe physico-mathématique de l'Académie Impériale des Sciences de St.-Pétersbourg*, [3] *2*, 170 (1860).—G.B.K.]

[64] In my reaction experiments (see my Beiträge [see p. 21 of reference 3]), I regarded these two black hydrates, which are easy to prepare, as reduced metals, an easily forgivable error. If the composition of these black hydrates is similar to that of the iridous and rhodous sesquioxides, then, surely, so are the hydrates which are richer in water.

[65] The possibility for this exists in the case of the rhodium base, for if its chlorine compound is treated with chlorine, then two equivalents of ammonia can thereby easily be decomposed and $3N\widehat{H_3R}h_2 + Cl_3$ be obtained. This method would be impractical for the iridium base, since it would obviously be converted into a quite different, iridium oxide-containing base, since the lower sesquichloride changes so readily to the higher chloride.

INTRODUCTION TO PAPER FIVE

[At about the time that Claus proposed his ammonia theory of metal-ammines, the concept of valence was being formulated and developed by a number of chemists—in particular, Kekulé, Frankland, Williamson, Odling, Kolbe, and Couper. During the late nineteenth and early twentieth centuries the principal difficulty in the field of valence was its application to *all* types of chemical compounds, and one of the main controversies involved whether or not a given element could possess more than one valence.[1,2] Inasmuch as coordination compounds pose a number of basic constitutional problems, it is not surprising that they became involved in the question of variable *vs.* constant valence.

What may come as a surprise to some readers, however, is that Kekulé's valence theory, which was so flexible and fruitful in the realm of organic chemistry, proved to be a virtual straitjacket when applied to inorganic compounds. Yet, by his own admission, Kekulé's concept of constant valence proved, in his own words, "embarrassing to the chemist." However, instead of abandoning this obviously untenable belief, he compounded his error by invoking a still more unsatisfactory concept in order to maintain it, namely, the concept of "molecular compounds,"[3] which he proposed in 1864.

Although Edward Frankland (1825–1899), Archibald Scott Couper (1831–1892), and other pioneers in the theory of valence readily admitted the possibility of variable valence, Friedrich August Kekulé (1829–1896), from his first statements[4] on the self-linking of carbon in 1858 until his death in 1896, adopted and rigidly adhered to the principle of constant valence. In spite of the mass of data that soon accumulated to contradict such a simple and admittedly attractive assumption, Kekulé dogmatically insisted that atomicity, which was the term that he used for valence, was, in his own words, "a fundamental property of the atom which is just as constant and unchangeable as the atomic weight itself."[5] The simplicity of this principle, however, was more than outweighed by the complicated and unrealistic formulas required in order to maintain it, and eventually the stubborn Kekulé stood virtually alone in its defense. Once again, as we have seen with Berzelius, the liberal of one generation had become the conservative of the next.

---

[1] C. A. Russell, *A History of Valency*, Leicester University Press, Leicester, England, 1971, Part III.
[2] G. B. Kauffman, *J. Chem. Educ.*, *49*, 813–817 (1972).
[3] A. Kekulé, *Compt. rend.*, *58*, 510 (1864).
[4] A. Kekulé, *Ann. Chem.*, *106*, 129 (1858).
[5] A. Kekulé, *Z. Chem.*, *7*, 689 (1864).

Kekulé's dichotomy of compounds into "atomic compounds" and "molecular compounds" was an attempt to buttress his theory of constant valence. According to Kekulé,

> Compounds in which all the elements are held together by the affinities of the atoms which mutually saturate one another could be called *atomic compounds*. They are the only ones which can exist in the vapor state. . . . We must distinguish a second category of compounds that I shall designate *molecular compounds*.[3]

A few examples should suffice to illustrate Kekulé's concept of "molecular compounds." Since Kekulé regarded the valences of nitrogen, phosphorus, and cobalt as invariably three, and of copper as invariably two, he was forced to consider phosphorus(V) chloride, ammonium chloride, copper(II) sulfate pentahydrate, and hexaammine-cobalt(III) chloride as "molecular compounds" with the formulas shown on the left-hand side of the following equations:

PCl_5 (Phosphorus(V) Chloride) $\qquad PCl_3 \cdot Cl_2 \xrightarrow{\Delta} PCl_3 + Cl_2$

NH_4Cl (Ammonium Chloride) $\qquad NH_3 \cdot HCl \xrightarrow{\Delta} NH_3 + HCl$

$CuSO_4 \cdot 5H_2O$ (Copper(II) Sulfate Pentahydrate) $\qquad CuSO_4 \cdot 5H_2O \xrightarrow{\Delta} CuSO_4 + 5H_2O$

$[Co(NH_3)_6]Cl_3$ (Hexaamminecobalt(III) Chloride) $\qquad CoCl_3 \cdot 6NH_3 \xrightarrow{\Delta}$ No Reaction

$CoCl_3 \cdot 6NH_3 \xrightarrow{NaOH} [Co(NH_3)_6](OH)_3$ (No $Co(OH)_3$ formed)

$CoCl_3 \cdot 6NH_3 \xrightarrow{H_2SO_4} [Co(NH_3)_6]_2(SO_4)_3$ (No $NH_4{}^+$ salt formed)

In a sense, Kekulé's concept of "molecular compounds" was a revival of Berzelius' dualistic theory whereby "secondary compounds" (in Kekulé's terminology, "atomic compounds") containing a small excess of electrical charge could still combine with other "secondary compounds" containing a small excess of opposite charge to form "tertiary compounds" (in Kekulé's terminology, "molecular compounds"). At most, Kekulé's artificial division of compounds into "atomic compounds," which obeyed the rules of classical valence theory, and into "molecular compounds," which did not obey these rules, had some limited value as a formal classification. However, in no way did it explain the nature or operation of the forces involved in the formation of "molecular compounds" except to assume that the forces were acting between molecules rather than between atoms.

Inasmuch as the forces acting between molecules were supposedly weaker than the forces acting between atoms, according to Kekulé, the resulting "molecular compounds" should be less stable than "atomic compounds." Indeed, some of the substances of limited thermal stability cited by Kekulé as prototypes of "molecular compounds," such as phosphorus(V) chloride, ammonium chloride, and copper(II) sulfate pentahydrate, did decompose in the vapor state. However, this thermal stability was a relative rather than an absolute phenomenon. For example, under certain conditions, such as the use of lower temperatures or the addition of decomposition products, many of these substances could be vaporized without decomposition. Therefore, many chemists began to regard Kekulé's classification as meaningless. In Mendeleev's words,

> Kekulé's division of chemical compounds into "atomic" and "molecular" types is artificial, arbitrary and unsound. . . . No practical test exists by which the two categories may be sharply separated.[6]

But Kekulé's stability criterion, or to be more accurate, instability criterion, failed completely in the case of many coordination compounds, especially the metal-ammines, which were classified as "molecular compounds" by sheer dint of necessity even though they were extremely resistant to heat and chemical reagents. For example, although hexaamminecobalt(III) chloride contains ammonia, it neither evolves this ammonia on mild heating nor does it react with acids to form ammonium salts. Also, despite its cobalt content, addition of a base to its aqueous solution fails to precipitate hydrated cobalt(III) hydroxide (p. 70).

Whereas Kekulé disposed of complex compounds by banishing them to the limbo of "molecular compounds," other chemists of the time developed highly elaborate theories in order to explain the constitution and properties of these colorful and intriguing substances. The most successful and widely accepted of such pre-Werner theories was undoubtedly the so-called chain theory, advanced in 1869 by Christian Wilhelm Blomstrand (1826–1897).

Living in Sweden during a transition period between the older and newer chemistry and being a scientific as well as a political conservative, Blomstrand sought to reconcile Berzelius' dualistic theory with the unitary and type theories. In fact, his best-known book, *Die Chemie der Jetztzeit* (1869), bears the subtitle "from the viewpoint of the electrochemical

---

[6] W. G. Palmer, *A History of the Concept of Valency to 1930*, Cambridge University Press, Cambridge, England, 1965, p. 82.

interpretation developed from Berzelius' theory." He considered the newer atomic theory to be "only a consequential development of Berzelius' atomic theory necessarily evoked by the force of many newly discovered facts."[7]

Despite his conservatism, Blomstrand was opposed to Kekulé's dogma of constant valence, and he strove to establish a sound and complete theory of variable valence. For him, Kekulé's dichotomy of compounds into *Valenzverbindungen*—those explicable by classical valence theory—and *Molekülverbindungen*—those that were not—was completely unacceptable. Blomstrand's assertion that "It has become the principal task of the newer chemistry to explain atomistically, i.e., from the saturation capacity [valence] of the elements, compounds which previously have been conceived of more or less definitely as molecular"[7] was chosen by no less an authority than Alfred Werner as the motto for his monumental *Neuere Anschauungen auf dem Gebiete der anorganischen Chemie.*

The latter half of the nineteenth century was a period of tremendous progress in the field of organic chemistry, and organic chemistry exerted a predominant influence on other branches of chemistry. Thus Blomstrand suggested that ammonia molecules could link together as —NH_3— chains, analogous to —CH_2—chains in hydrocarbons. These chains involved "quinquevalent" nitrogen, an idea that Kekulé found anathema. The number of ammonia molecules associated with the metal, i.e., the length of the chain, depended upon the metal and its valence. This point was later accounted for more adequately by Werner's concept of the coordination number. Blomstrand made provision for different reactivities of various atoms and groups. For example, halogen atoms that could not be precipitated immediately by silver nitrate were called "nearer" and were considered to be bonded directly to the metal atom. Halogen atoms that could be precipitated immediately by silver nitrate were called "farther" and were considered to be bonded through the ammonia chains. These two different kinds of bonding were later explained more satisfactorily by Werner's terms "nonionogenic" and "ionogenic," respectively, and by his concepts of inner and outer spheres of coordination. Despite the chain theory's admitted limitations, it permitted the correlation of a considerable amount of empirical data.

In 1883 Blomstrand's friend Sophus Mads Jørgensen (1837–1914)[8] demonstrated that tertiary amines, which contain no hydrogen, are capable of forming compounds that are completely analogous to the

---

[7] C. W. Blomstrand, *Die Chemie der Jetztzeit vom Standpunkte der electrochemischen Auffassung aus Berzelius' Lehre entwickelt*, Carl Winter's Universitätsbuchhandlung, Heidelberg, 1869. Blomstrand's first ideas on his chain theory of coordination compounds appear here, beginning on p. 280.

[8] S. M. Jørgensen, *J. prakt. Chem.*, [2] *27*, 478–486 (1883).

metal-ammines both in their composition and their properties. Graham's ammonium theory (Paper Three) conceived of metal-ammines as salts in which some of the hydrogen atoms of the ammonium group were replaced by metal atoms. Since tertiary amines contain no hydrogen, Jørgensen's discovery effectively eliminated the ammonium theory from serious consideration as an explanation for the constitution of the metal-ammines. Therefore, chemists at that time were forced to assume the existence of ammonia chains copied from hydrocarbons or to conceive of metal-ammines as "molecular compounds." Since Kekulé's theory really explained nothing and only "substituted a beautiful word for a confused concept," to use Werner's words, the chain theory (Paper Five),[9] as modified and developed by Jørgensen (Paper Six),[10] became the most popular and satisfactory way of accounting for metal-ammines. It held sway for roughly a quarter of a century until it was displaced by Werner's coordination theory.

Christian Wilhelm Blomstrand[11] was born on October 20, 1826, in the small town of Växjö in the central Swedish province of Småland, where his father was a teacher at the local *Gymnasium* (high school). In 1844 he entered the University of Lund where he studied mineralogy. His interest in chemistry began only after he had been awarded his doctorate in 1850 and had been named the first recipient of the Berzelius scholarship. He then began to study chemistry and became an assistant to Professor of Chemistry Nils Johannes Berlin (1812–1900), who had been a student of Berzelius'. In 1854 he became a *Privat-Dozent*; his *Habilitationsschrift* dealt with bromine and iodine compounds of tin.

Inasmuch as the outlook for academic advancement at Lund was not favorable, Blomstrand accepted a position as instructor in the technical school at Malmö. He remained at this post only a short time, for in 1855 Johann Bernhard von Borck, the *adjunct* in chemistry at Lund, died unexpectedly, and the following year Blomstrand was named as his successor. In 1862 Prof. Berlin transferred to the Medical Faculty, and Blomstrand became Professor of Chemistry and Mineralogy, a position which he occupied until his retirement in 1895. He continued his scientific work, both experimental and literary, however, until his sudden death at Lund on November 5, 1897. He never married, and except for a brief

---

[9] C. W. Blomstrand, *Ber.*, *4*, 40–52 (1871).

[10] S. M. Jørgensen, *Z. anorg. Chem.*, *19*, 109–157 (1899).

[11] The entire September (No. 9) issue (pp. 234–314) of *Svensk Kemisk Tidskrift*, Vol. *38*, 1926, is devoted to articles on various aspects of Blomstrand's career written by F. Elander, E. Wrangel, L. Smith, K. A. Grönwall, and L. Ramberg. E. von Meyer, *J. prakt. Chem.*, [2] *56*, 397–400 (1897); P. Klason, *Ber.*, *30*, 3227–3241 (1897); G. B. Kauffman, *Dictionary of Scientific Biography*, ed. by C. C. Gillispie, Charles Scribner's Sons, New York, 1970, Vol. 2, pp. 199–200; *Ann. Sci.*, *32*, 13–37 (1975).

period in 1861 when he served as mineralogist on a polar expedition to Spitzbergen (Svalbard), Norway, and a few other mineralogical trips, he remained at Lund.

Blomstrand's early preoccupation with the earth acids (Group VB elements) remained with him, and his experimental work included studies of the compounds of niobium, tantalum, molybdenum, and tungsten, especially their halides and oxyhalides.[12] He also prepared heteropoly acids of iodic and periodic acids with chromic, molybdic, and tungstic acids.[13] His discovery that the "addition compounds" of platinous chloride and alkyl sulfides occurred in two isomeric forms (*cis*- and *trans*-$[Pt(R_2S)_2Cl_2]$) was a valuable contribution to the stereochemistry of coordination compounds.[14] In addition to strictly chemical work, he characterized and analyzed many minerals, especially those of the rarer elements or those of unknown composition, such as monazite, ilmenite, tantalite, niobite, and euxenite.[15] Most of Blomstrand's theoretical works are polemical, but more often conciliatory than inflammatory. For example, with his azoammonium theory[16] he attempted to reconcile opposing views of diazonium compounds by Hantzsch, Bamberger, and others.—G.B.K.]

---

[12] C. W. Blomstrand, *J. prakt. Chem.*, *82*, 408, 433 (1861); *ibid.*, *89*, 230 (1863).
[13] C. W. Blomstrand, *J. prakt. Chem.*, [2] *40*, 305–340 (1889); *Z. anorg. Chem.*, *1*, 10–50 (1892).
[14] C. W. Blomstrand, *J. prakt. Chem.*, [2] *38*, 345–365, 498–530 (1888).
[15] C. W. Blomstrand, *Geol. Fören. i Stockholm Förh.*, *9*, 160 (1887); *ibid.*, 379 (1889); *J. prakt. Chem.*, [2] *41*, 266–277 (1890); K. A. Grönwall, *Svensk Kem. Tidskr.*, *38*, 272–276 (1926).
[16] C. W. Blomstrand, *Ber.*, *8*, 51–55 (1875); *ibid.*, *30*, 1265 (1897); *J. prakt. Chem.*, [2] *53*, 169–197 (1896); *ibid.*, [2] *54*, 305–339 (1896); *ibid.*, [2] *55*, 481–501 (1897); L. Smith, *Svensk Kem. Tidskr.*, *38*, 300–314 (1926).

5

Christian Wilhelm Blomstrand

Metal-Ammonias or Metal-Ammines

[Translation of "Ueber die Metallammoniake oder die Metallammine," *Berichte der Deutschen Chemischen Gesellschaft*, *4*, 40–52 (1871).]

(Received on January 13.)

The real subject of my first article published in these Berichte[1]: "Toward an Understanding of Copulated Compounds in Inorganic Chemistry" [*Zur Kenntniss der gepaarten Verbindungen der anorganischen Chemie*] was only some factual support for my earlier expressed view[2] that the cyanide double salts and the double and triple salts of nitrous acid, just like and corresponding completely to metal-ammonias [*Metallammoniake*], are copulated compounds [*gepaarte Verbindungen*] of quinqueatomic[3] nitrogen [*fünfatomiger Stickstoff*],[4] in whose existence, along with triatomic[3] nitrogen, the first condition of these surprisingly stable compounds should be sought. For example, we may compare the NH^3—, NC—, and NO— compounds of platinum and rhodium, of nickel and cobalt, and so forth.

I now take the liberty of expounding somewhat more completely than the earlier limited space permitted me on my view of the metal-

[1] [*Berichte, 2,*] 202[–206] (1869).

[2] [*Die*] *Chemie der Jetztzeit* [*vom Standpunkte der electrochemischen Auffassung aus Berzelius Lehre entwickelt,* Carl Winter's Universitätsbuchhandlung, Heidelberg, 1869], pp. 35, 280, 303, 350.

[3] ["Atomicity" [*Atomizität*] was the term favored by Kekulé to denote what we now call "valence" (F. A. Kekulé, *Bull. Acad. roy. Belg.*, [2] *10*, 71 (1860); *Lehrbuch der organischen Chemie*, Ferdinand Enke, Erlangen, 1861). Hence the German suffix "-atomig," rendered in this paper as "-atomic," is equivalent to our modern suffix "-valent."—G.B.K.]

[4] [Nitrogen, as predicted by modern orbital theory, cannot exceed a covalence of four, yet for many years the concept of quinquevalent (pentavalent) nitrogen was widely used to explain the constitution of ammonium salts and metal-ammines (G. B. Kauffman, *Isis*, *64*, 78–95 (1973)). According to Blomstrand (footnote 2, p. 113), the concept was first proposed by Heinrich Franz Peter Limpricht (1827–1909).—G.B.K.]

ammonia bases, which thus are also decisive for the negative cyanide and nitrosyl complexes. The interesting investigations of P. T. Cleve,[5] upon which I report in the following correspondence, are the immediate cause for this, aside from the fact that in recent times a number of events have occurred which could cause me to re-investigate these theoretically not unimportant questions. I especially took into account here the views of Dr. Odling[6] reported by [R.] Gerstl in these *Berichte* (III, pp. 682 [–687 (1870)]) "on the Platinum Compounds of Ammonia" ["über die Platinverbindungen des Ammoniaks"], since, of the previously presented views of this much-discussed matter, they, in any case, come closest to mine.

Of course, I could refer to the above-cited work (p. 280 ff.)[7] in regard to my general reasoning; however, since I cannot presuppose any greater circulation of this work, I must here relate anew the more essential points of my line of reasoning.

As is well known, we were led rather early (1838) to assume so-called *copulated compounds* [*gepaarte Verbindungen*][8] in addition to the halide [*Haloid*] and amphide [*Amphid*] compounds of simple and complex radicals known since ancient times from inorganic chemistry.

Since, by the use of mixed and condensed types, it became possible for the more recent theory of types[9] to formulate typically even the more complicated compounds, it was found unnecessary to

---

[5] [Per Theodor Cleve (Stockholm, Sweden, Feb. 10, 1840—Uppsala, Sweden, June 18, 1905), Professor of Chemistry at Uppsala University, carried out much research on the platinum-ammines and the rare earths (he discovered holmium and thulium).—G.B.K.]

[6] [William Odling (Southwark (London), England, Sept. 5, 1829—Oxford, England, Feb. 17, 1921), Professor of Chemistry at the Royal Institution and later at Oxford University, is best known for his formulas for ozone and bleaching powder, but he also made fundamental pioneering contributions to valence and the periodic law.—G.B.K.]

[7] [See footnote 2.—G.B.K.]

[8] [The term "copula" was originally used by Charles Gerhardt (*Ann. chim. phys.*, [2] *72*, 184 (1839)) to indicate the organic part of a molecule that might be joined to an inorganic part. In what was one of the earliest systematic attempts to explain the constitution of coordination compounds Berzelius (1841) applied these ideas to metal-ammines, viewing them as copulated compounds consisting of ammonia and a copula. See also p. 47.—G.B.K.]

[9] [Largely as a result of his research on organic substitution (*Ann. chim. phys.*, [2] *56*, 113, 140 (1834)), J. B. A. Dumas (*Ann.*, *33*, 179, 259 (1840)) proposed his theory of types postulating that "in organic chemistry there are certain types which remain unchanged, even when their hydrogen is replaced by an equal volume of chlorine, bromine, and iodine." The theory was primarily an

take this difference into account. Thus, for example, we read in Kekulé's[10] Textbook[11] (I, p. 209): "All efforts to give a new meaning to the word 'copulated' ['*gepaart*']—introduced twenty years ago, of which one became gradually convinced that it actually had no meaning and that the differences which it was intended to express do not in fact occur—all these efforts have remained fruitless and will most probably remain so since such differences do not exist." It was found especially suitable to emphasize the meaninglessness of Berzelius'[12] views on copulation [*Paarung*].[13]

The assumption of bonding by copulation was not a result of chemical speculation. Newly discovered facts (benzenesulfonic acid[14] and so forth), from whose explanation the entire new point of view has gradually developed, absolutely necessitated it.

If we adhere to the concept of copulation as Berzelius once formulated it, the following comparison between the old and new copulated formulas is offered.

Berzelius wrote as copulated substances:[15]

Acetic acid: $\bar{\mathrm{H}}\mathrm{O},\ \mathrm{C^2O^3 \cdot C^2\bar{H}^3}$[16]
Aniline chloride: $\mathrm{C^{12}\bar{H}^4 \cdot NH^4 \cdot \bar{C}l}$.[17]

---

attempt to explain the phenomena of organic chemistry that were not explicable on the basis of Berzelius' dualistic electrochemical theory. As developed by Auguste Laurent, Charles Gerhardt, A. W. Hofmann, and A. W. Williamson, it eventually led to the concept of valency. See O. T. Benfey, *Classics in the Theory of Chemical Combination,* Dover Publications, Inc., New York, 1963.—G.B.K.]

[10] [(Friedrich) August Kekulé (Darmstadt, Germany, Sept. 7, 1829—Bonn, Germany, July 13, 1896), Professor of Chemistry at the University of Ghent and later at the University of Bonn, was a major figure in the development of structural organic chemistry and valency.—G.B.K.]

[11] [*Lehrbuch der organischen Chemie,* Ferdinand Enke, Erlangen, 1861.—G.B.K.]

[12] [Jöns Jacob Berzelius (Väversunda, Sweden, Aug. 20, 1779—Stockholm, Sweden, Aug. 7, 1848) was one of the major figures and supreme arbiter of chemistry during the first half of the nineteenth century. Blomstrand attempted to reconcile Berzelius' dualistic theory with the newer unitary and type theories. —G.B.K.]

[13] [For example, Laurent wrote, "What then is copula? A copula is an imaginary body, the presence of which disguises all the chemical properties of the compounds with which it is united" (*Chemical Method,* Cavendish Society, London, 1855, pp. 37, 204).—G.B.K.]

[14] [Discovered by E. Mitscherlich (*Ann. Phys., 32,* 227 (1834)).—G.B.K.]

[15] [Barred atoms were introduced by Berzelius to denote doubled atoms or two atoms, e.g., $\bar{\mathrm{H}}\mathrm{O} = \mathrm{H_2O}$ (*Jahresbericht, 7,* 72 (1828)).—G.B.K.]

[16] [In modern notation, CH_3COOH.—G.B.K.]

[17] [Aniline hydrochloride, $C_6H_5NH_2 \cdot HCl$, or better, $[C_6H_5NH_3]Cl$. The original incorrectly reads "$\mathrm{C^{12}\bar{H}^4 \cdot NH^4 \cdot Cl}$."—G.B.K.]

On the other hand, we first write:

$$\not{\mathrm{H}}\mathrm{O}, \overset{\mathrm{IV}}{\mathrm{C}}{}^2\mathrm{O}^2 \cdot \overset{\mathrm{IV}}{\mathrm{C}}{}^2\not{\mathrm{H}}^3 \cdot \mathrm{O} \quad \text{and} \quad \mathrm{C}^{12}\not{\mathrm{H}}^5 \cdot \overset{\mathrm{V}}{\not{\mathrm{N}}}\not{\mathrm{H}}^3 \cdot \not{\mathrm{C}}\mathrm{l},$$

but more correctly we write:

$$\mathrm{H} \cdot \mathrm{O} \cdot \overset{\mathrm{IV}}{\mathrm{C}}\mathrm{O} \cdot \mathrm{CH}^3 \quad \text{and} \quad \mathrm{C}^6\mathrm{H}^5 \cdot \mathrm{NH}^3 \cdot \mathrm{Cl}$$

with freely acting atoms of the hydrogen group actually behaving atomistically.

Berzelius therefore saw in copulated formulas a very great gain for science, exactly because the separation of the chemically active constituent from the inactive constituent that was completely inactive chemically (the so-called copula), which was thereby intended, made possible a much deeper insight into the inner arrangement of the atoms than had the previous mere separation of the radicals. We boast of having come closer to the general laws of arrangement, and we universally speak of atoms bound to one another rather than, as previously, of molecules thrown at one another [*aneinander geworfen*]. Just as C^6H^5 and Cl are now added onto NH^3, so were $C^{12}\not{H}^4$ and $\not{C}l$ previously added onto $\not{N}\not{H}^4$ in exactly the same manner. The difference concerned only the quite different chemical (or, as one could say, electrochemical) mode of action of the two members [*Membra*]. I believe that even now it will hardly be claimed that in aniline chloride C^6H^5 and Cl act chemically in the same manner.

Of course, the term "copulation" [*Paarung*] can very easily be avoided. The concept, however, still remains.

Reiset's[18] platinum base[19] contributed more than anything else to the development of Berzelius' theory of copulation. He gave it the formula

$$\mathrm{Pt}\not{\mathrm{N}}\not{\mathrm{H}}^2 \cdot \not{\mathrm{N}}\not{\mathrm{H}}^4\not{\mathrm{C}}\mathrm{l}.$$

If here too we adhere to the chemically active ammonium, then, according to the laws of atomicity,[3] this formula will be altered to:

$$\overset{\mathrm{I}}{\mathrm{Pt}} \cdot \overset{\mathrm{V}}{\not{\mathrm{N}}}\not{\mathrm{H}}^3 \cdot \overset{\mathrm{V}}{\not{\mathrm{N}}}\not{\mathrm{H}}^3 \cdot \not{\mathrm{C}}\mathrm{l}$$

or with resolved atoms:

$$\overset{\mathrm{II}}{\mathrm{Pt}}\begin{matrix} \overset{\mathrm{V}}{\mathrm{N}}\mathrm{H}^3 \cdot \overset{\mathrm{V}}{\mathrm{N}}\mathrm{H}^3 \cdot \mathrm{Cl} \\ \overset{\mathrm{V}}{\mathrm{N}}\mathrm{H}^3 \cdot \overset{\mathrm{V}}{\mathrm{N}}\mathrm{H}^3 \cdot \mathrm{Cl} \end{matrix}$$

---

[18] [Jules Reiset (1818–1896).—G.B.K.]

[19] [Reiset's so-called first chloride. In modern notation, tetraammineplatinum(II) chloride, $[Pt(NH_3)_4]Cl_2$.—G.B.K.]

If we still wish to hear of a theory of copulation, then it may be said: it is completely encompassed in this formula for Reiset's first base.

Thus we can call compounds of polyatomic[3] *elements, which act chemically as a unified whole, copulated compounds.*

$\overset{\text{IV}}{\text{Co}}\cdot\overset{\text{IV}}{\text{Co}}$, $\overset{\text{IV}}{\text{Pt}}\cdot\overset{\text{IV}}{\text{Pt}}$, $\overset{\text{IV}}{\text{C}}\cdot\overset{\text{VI}}{\text{S}}$, $\overset{\text{IV}}{\text{C}}\cdot\overset{\text{IV}}{\text{C}}$, $\overset{\text{IV}}{\text{C}}\cdot\overset{\text{IV}}{\text{C}}\cdot\overset{\text{IV}}{\text{C}}$, and so forth, are examples of these copulated substances in the broader meaning of the word. If CH^3 combines with CH^3, we say that there is formed an ethyl compound, in which not C but C^2 is the radical base [*Grundlage*].

Moreover, it is in the nature of the matter itself that here too the differences are only relative. Thus, for example, triatomic[3] nitrogen forms a natural transition between monatomic[3] Cl, which definitely acts as a radical, and quinqueatomic[3] nitrogen, which forms radicals, especially with H and O.

In order to clarify somewhat the laws of copulation, as conceived in this manner, we cannot refrain from making a close connection between it and the *change in atomic valence* [*Wechsel des Atomwerthes*],[20] just as, conversely, if the possibility of such a phenomenon is denied, the assumption of the bonding in question for many elements is impossible to begin with.

Experience shows us that in the two main groups of elements, hydrogen and oxygen, the increase in atomic valence proceeds regularly in even numbers, as in $\overset{\text{I}}{\text{Cl}}$, $\overset{\text{III}}{\text{Cl}}$, $\overset{\text{V}}{\text{Cl}}$, $\overset{\text{VII}}{\text{Cl}}$; in $\overset{\text{III}}{\text{N}}$ and $\overset{\text{V}}{\text{N}}$; $\overset{\text{II}}{\text{C}}$ and $\overset{\text{IV}}{\text{C}}$; $\overset{\text{II}}{\text{S}}$, $\overset{\text{IV}}{\text{S}}$, and $\overset{\text{VI}}{\text{S}}$; $\overset{\text{II}}{\text{Pt}}$ and $\overset{\text{IV}}{\text{Pt}}$; $\overset{\text{II}}{\text{Co}}$ and $\overset{\text{IV}}{\text{Co}}$, and so forth.

The presence of this regular change in itself makes it very likely that the points of attack of, for example, a diatomic[3] or triatomic[3] element cannot be of the same type as those which on increasing to tetraatomicity[3] and quinqueatomicity[3] become active anew.

These latter could be called *extraradical points of attack* [*extra-radikale Angriffspunkte*] in view of the irrefutable experience that the elements act generally as radicals only in the lower stages of combination, and the atomicity[3] of carbon and nitrogen, for example, could be expressed in the following manner:

| · · | | · · · |

---

[20] [Although many of the pioneers in the theory of valency readily admitted the possibility of variable valency, Kekulé, from his first statements on the self-linking of carbon in 1858 until his death in 1896, adopted and rigidly adhered to the doctrine of constant valency.—G.B.K.]

On mutual bonding of two atoms, the complex acting as a whole was assigned the corresponding graphic expression:

If we similarly imagine three or more atoms arranged together in a row, then, in accordance with this notation, we obtain everywhere points of attack which act outward, whether extraradical or belonging to tetraatomic[3] and quinqueatomic[3] elements.

We call those carbon compounds such as $H\cdot CH^2\cdot H$, $H\cdot CH^2\cdot CH^2\cdot H$, $H\cdot CH^2\cdot CH^2\cdot CH^2\cdot H$,[21] which according to all appearances are constructed according to this scheme, *normal* in preference to other compounds.

Of course, only experience can tell us whether, and to what extent, each particular element is capable of polymerizing, so to speak, in this manner. Thus, with most tetraatomic[3] elements, bonding seems to stop with two atoms, and moreover at times it seems to occur with special ease as with Fe, Co, Rh, and at times with more or less difficulty as with Pb, Sn, Mn, Pt, and so forth. With no element is this capacity for self-bonding as unlimitedly prevalent as with carbon.

Among other elements, as with quinqueatomic[3] nitrogen, which actually is our subject here, this capacity seems subject to certain external conditions, for with at least extremely rare exceptions, it here seems to presuppose the bondedness [*Gebundensein*] of the element to a metal which serves to some extent as a radical base [*Grundlage*] of the entire molecule. Thus the metal-ammonias[22] with the general graphic expression:

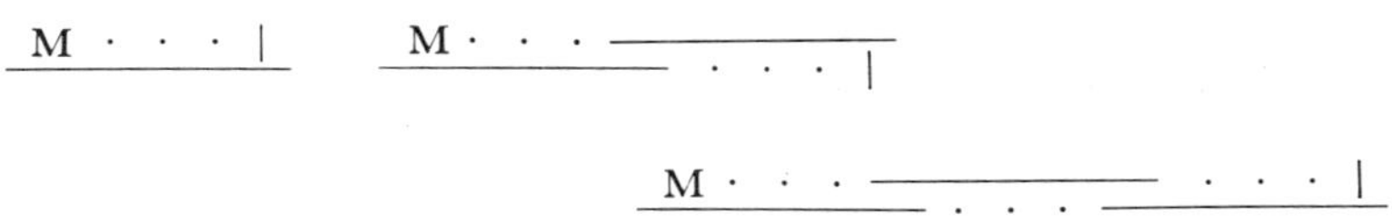

would be the only examples of this type of nitrogen complex that is known with certainty.

If the metal is replaced by an organic radical, that is, by carbon, the ammonia never seems to be doubled. Only simple ammonia compounds are formed, which, on the other hand, seem to be less

[21] [CH_4, CH_3CH_3, $CH_3CH_2CH_3$.—G.B.K.]

[22] [In modern terminology, metal-ammines.—G.B.K.]

normal among the metals. It is also noteworthy that among the metal-ammonias the ammonium which acts outward is far more difficult to convert into free-acting triatomic[3] ammonia. To the best of my knowledge, the few examples of this have been made known to us by Cleve (cf. the Correspondence).[23]

The particular importance of the metal as the first radical base of the entire positive complex is also expressed very clearly by the fact that the nature of the metal apparently determines not only the more or less firm coherence of the entire complex (cf., for example, on the one hand Pt and Co and, on the other hand, Ag, Cu, Zn, and Ni) but also the highest number of nitrogen atoms which can occasionally combine and as a rule unite;[24] cf., for example, Cu and Pt, in which at most two nitrogen atoms unite, Ni, Co, Ir, and Rh, which as a rule stimulate three nitrogen atoms to reciprocal bonding, as in

$$\mathrm{Pt}\begin{matrix}\mathrm{NH^3\cdot NH^3\cdot Cl}\\\mathrm{NH^3\cdot NH^3\cdot Cl}\end{matrix}\qquad\mathrm{Co}\begin{matrix}\mathrm{NH^3\cdot NH^3\cdot NH^3\cdot Cl}\\\mathrm{NH^3\cdot NH^3\cdot NH^3\cdot Cl.}\end{matrix}$$

That meanwhile in this kind of reciprocal bonding, *intraradical points of attack* can also be active, according to our view, already follows from the fact that certain elements can appear at the lowest stage of combination with two or more atoms acting jointly, such as $\overset{\mathrm{I}}{\mathrm{Hg}}\cdot\overset{\mathrm{I}}{\mathrm{Hg}}$ in mercurous chloride according to the usual view, as sulfur in the polysulfides of the radicals and in the polythionic acids, and so forth. The double bonding of carbon in the aromatic series, of triatomic[3] nitrogen in azobenzene and related substances, of triatomic[3] and quinqueatomic[3] nitrogen in azoxybenzene, as well as, according to my view, of chlorine in halide double salts and of quinqueatomic[3] nitrogen in cyanide and nitroxyl double salts would thus be examples of this more or less deviant but certainly not rare form of bonding, to be expressed graphically in somewhat the following manner

[23] [In the *Berichte der Deutschen Chemischen Gesellschaft*.—G.B.K.]

[24] [This fact was later accounted for by Alfred Werner in his concept of the coordination number (*Z. anorg. Chem.*, *3*, 267–330 (1893)); for a discussion and English translation see G. B. Kauffman, *Classics in Coordination Chemistry, Part I: The Selected Papers of Alfred Werner*, Dover Publications, Inc., New York, 1968, pp. 5–88.—G.B.K.]

Among these belong, for example:

$$\begin{matrix} K \cdot NC \cdot NC \\ K \cdot NC \cdot NC \end{matrix} Pt \quad \text{and} \quad \begin{matrix} K^2 \cdot NC \cdot NC \cdot NC \\ K^2 \cdot NC \cdot NC \cdot NC \end{matrix} Co$$

$$\begin{matrix} K \cdot O \cdot NO \cdot NO \cdot O \\ K \cdot O \cdot NO \cdot NO \cdot O \end{matrix} Pt \quad \text{and} \quad \begin{matrix} K^2 \cdot O^2 \cdot NO \cdot NO \cdot NO \cdot O \\ K^2 \cdot O^2 \cdot NO \cdot NO \cdot NO \cdot O \end{matrix} Co,[25] \quad \text{and so forth.}$$

It follows automatically that among combinations of this type it must often be very difficult to decide with any certainty about the exact type of the bonding, because not rarely the possibility exists for assuming lower combining values instead of double or single bonding. For example, trimolybdenum in chloromolybdenum chloride $Mo^3Cl^4 \cdot Cl^2$ could be represented graphically either as

 or ,[26]

a perhaps hexaatomic[3] Pt^4 as either

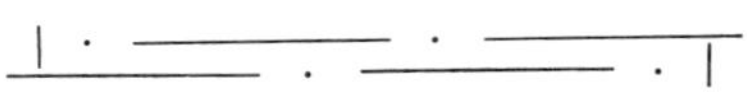 or 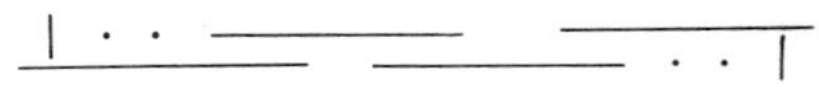

and so forth. We know that in organic chemistry such difficult-to-decide questions repeatedly occur (for example, fumaric acid,[27] cinnamic acid,[28] and so forth).

In this connection we may also mention the fact that at times intraradical points of attack can also be active even with single bonding. That this occurs frequently with carbon radicals is sufficiently well known. The numerous so-called *iso substances* are examples of this. That carbon radicals can likewise enter into the interior of the ammonia, so to speak, has also been more than adequately proven. On the other hand, it is very doubtful whether nitrogen acting quinqueatomically[3] can also similarly bind nitrogen.

It is necessary for us to compare with each other here our so-called extraradical points of attack, that is, to compare here first of all carbon and nitrogen in terms of their possibly different mode of acting.

---

[25] $[K_2[Pt(CN)_4]$, $K_4[Co(CN)_6]$, $K_2[Pt(NO_2)_4]$, and $K_4[Co(NO_2)_6]$, respectively.—G.B.K.]

[26] [Since these two diagrams are identical, one of them is obviously incorrect.—G.B.K.]

[27] $\left[\begin{matrix} HOOC \cdot CH & \\ \quad \| & \\ & HC \cdot COOH \end{matrix}\right]$.—G.B.K.]

[28] $\left[\begin{matrix} C_6H_5 \cdot CH \\ \quad \| \\ HOOC \cdot CH \end{matrix}\right]$.—G.B.K.]

$H \cdot CH^2 \cdot H$ is formed very readily and is very stable; the same is true of $H \cdot CH^2 \cdot CH^2 \cdot H$, and so forth. The chlorine-containing substances $H \cdot CH^2 \cdot Cl$ and $H \cdot CH^2 \cdot CH^2 \cdot Cl$, and so forth, are somewhat difficult to decompose by double exchange,[29] but yet they behave in every respect like chlorine compounds of independent radicals CH^3 and C^2H^5.

Nitrogen hydride [*Stickstoffwasserstoff*] $H \cdot NH^3 \cdot H$, not to mention higher complexes, does not exist. In the substance $H \cdot NH^3 \cdot Cl$,[30] NH^4 so definitely shows the properties of a radical that in this respect it vies with the true alkali metals. These facts could be expressed thus: extraradical points of attack behave in an electrochemically different manner with all polyatomic radical-forming elements, but with nitrogen far more than with carbon.

It seems to me to be a natural consequence of this, that whereas methyl quite normally replaces the hydrogen in methyl CH^3 or in methylene CH^2, a replacement of the hydrogen of ammonia NH^3 or ammonium NH^4 by ammonium must be impossible from the start. The fifth point of attack of the quinqueatomic[3] nitrogen bound to the hydrogen takes a definitely negative position.

I see in this the intrinsic defect in the many older formulas for Reiset's First Base,[19] according to which ammonium is assumed to be a substituent for hydrogen in ammonium, as, for example, has been rather generally assumed for some time, according to Kolbe:[31]

$$\left.\begin{array}{l} \cancel{H}^2 \\ \cancel{H}^4\cancel{N} \\ \cancel{Pt} \end{array}\right\} \cancel{N} \cdot \cancel{C}l.$$

From a typical viewpoint, the Gerhardt[32] substitution formula $\left.\begin{array}{l} \cancel{H}^5 \\ Pt \end{array}\right\} \cancel{N}^2, \cancel{H}\cancel{C}l$ was interpreted completely empirically and had nothing at all to do with the inner position of the atoms.

---

[29] [That is, metathetical substitution reactions.—G.B.K.]

[30] [Ammonium chloride, NH_4Cl.—G.B.K.]

[31] [Hermann Kolbe (Elliehausen bei Göttingen, Germany, Sept. 27, 1818—Leipzig, Nov. 25, 1884), Professor of Chemistry at the Universities of Marburg and Leipzig, was probably the most controversial figure in chemistry at the time. He was particularly known for his critical and polemical writings.—G.B.K.]

[32] [Charles Frédéric Gerhardt (Strasbourg (Alsace), France, Aug. 21, 1816—Strasbourg, Aug. 19, 1856) originated the term copula (see footnote 8).—G.B.K.]

Claus'[33] formula $2\mathrm{NH}^3 \cdot \mathrm{PtCl}$, which also met with much approval, was taken as a purely empirical formula, in any case, as the simplest, but it became defective as soon as it was intended to function as a rational copulation formula. It suffered from the all too often occurring, unclear conception of the concept of copulation; the entire ammonia content was a chemically inactive copula.

The formulas which have been proposed since the development of the strictly atomistic conception are usually faulty in the same way as the older substitution formulas just discussed. Thus, for example, Frankland (*Lecture Notes for Chemical Students*)[34] and H. L. Buff (*Lehrbuch*)[35] write:

$$\mathrm{Pt}\begin{array}{l}\mathrm{NH^2(NH^4)Cl}\\ \mathrm{NH^2(NH^4)Cl.}\end{array}$$

Ammonium is substituted for hydrogen in ammonium; platinum and chlorine are bound to the same nitrogen atom.

More correct, of course, is the formula which Kolbe has recently given (*Journal für praktische Chemie*, [[*2*]*2*], 220 (1870)) or

$$\left.\begin{array}{r}\mathrm{Pt}\left\{\left[\begin{array}{l}\mathrm{H^3N}\\ \mathrm{H^3N}\end{array}\right]\right.\\ \mathrm{H^2}\\ \mathrm{H^2}\\ \mathrm{H^2}\end{array}\right\}\mathrm{N^2Cl^2},$$ according to which "the radical $\mathrm{Pt}\begin{array}{l}\mathrm{H^3N}\\ \mathrm{H^3N}\end{array}$ takes the place of two hydrogen atoms in one diammoniumdichloride." However, I see no reason for preferring these typical substitution formulas to the true arrangement formulas [*Lagerungsformeln*]. In any case, they give a very unclear idea of the arrangement thought possible, aside from the fact that they unnecessarily take up too

[33] [Carl Ernst Claus, known also by his Russian name Karl Karlovich Klaus (Dorpat, Estonia (now Tartu in the Estonian SSR), Jan. 23 (Jan. 11, O.S.), 1796—Dorpat, Mar. 24 (Mar. 12, O.S.), 1864), Professor of Chemistry at the University of Kazan and later Professor of Pharmacy at the University of Dorpat, is best known as the discoverer of ruthenium. For his theory of metal-ammines see Paper Four.—G.B.K.]

[34] [Edward Frankland (Churchtown, England, Jan. 18, 1825—Golaa, Norway, Aug. 8, 1899) is best known for his discovery of organometallic compounds and his role in the concept of valency. *Lecture Notes for Chemical Students; Embracing Mineral and Organic Chemistry*, J. Van Voorst, London, 1866.—G.B.K.]

[35] [Heinrich L. Buff (Rödelheim (Wetterau), Germany, May 23, 1805—Giessen, Germany, Dec. 23, 1878), Professor of Physics at the University of Giessen. *Kurzes Lehrbuch der anorganischen Chemie entsprechend den neueren Ansichten*, Ferdinand Enke, Erlangen, 1868.—G.B.K.]

much space. Similarly written formulas for the cobalt bases, even for the simple ones (if I understand it correctly, as the dichloride of a tetraammonium) would be almost inconceivable atomistically.

It may be mentioned in passing that the above-mentioned authors also do not give an atomistic formula for the cobalt bases. Odling, to whose views we now turn, also does not consider these substances.

As has already been noted, the formulas proposed by Odling come closest to mine and, so to speak, coincide with them, for the ammonia bases are compared with ethylene, and thus the basic principles, also, which I developed above, are correctly recognized, at least in regard to the main point.

Yet I take the liberty of adding some incidental remarks regarding the detailed development of Odling's view.

The comparison between $H^2N \cdot H^3N$ or $H^5 \cdot N^2$ as[35a] "monovalent radical ammonamidogen" with the monovalent radical methylmethylene or ethyl $H^3C \cdot H^2C$ or H^5C^2 can very easily lead to completely erroneous ideas. In any case, H^5N^2 does not function as a monovalent radical in the same sense as H^5C^2. $H^3N \cdot NH^3$ corresponds to $H^2C \cdot CH^2$. It is superfluous to proceed further because in both cases the remaining atomic units act quite differently. Moreover, it is noteworthy that although the author initially attempts to prove that one cannot avoid viewing ammonium chloride as NH^4Cl and not as NH^3HCl,[36] still in all cases in which a double ammonia is involved, he assumes that the chemically active nitrogen is triatomic,[3] although it is exactly here that the quinqueatomicity[3] is so prominent that alkalies almost never split off ammonia but instead split off ammonium oxide hydrate. Typical considerations need not cause this, since H^5N^2 is indeed unconditionally assumed to be quinqueatomic[3] nitrogen.

Finally, I cannot refrain from noting that it seems to me that the author does not adhere with sufficient rigor to the first principle of the atomistic conception, that is, to the combining values which are actually given. Thus at first glance, it can hardly be misunderstood that the compounds designated as ammonodiplatinamines can in no way be diplatinum compounds. The substances in question are octaatomic.[3] Diplatinum can at most be assumed to be

---

[35a] [The original incorrectly reads "$H^5 \cdot N$."—G.B.K.]

[36] [In order to maintain a constant valency of three for nitrogen, Kekulé considered ammonium chloride as a "molecular compound" (*Molekülverbindung*) of ammonia and hydrogen chloride, $NH_3 \cdot HCl$.—G.B.K.]

hexaatomic,[3] as long as we have as yet no reason for assuming the singly acting platinum to be more than tetraatomic.[3]

In regard to the experimental proofs, it would be desirable if, as is evident from the historical introduction, the investigations carried out in Sweden had not remained completely unknown, which is all the more remarkable because the most important of the works of Cleve, which are in all respects noteworthy, that is, those on Reiset's First Base and Gros' Base,[37] may be found summarized in the *Jahresbericht* (*1867*, 321) and, for example, *Bull. Soc. Chim.* (*7*, 12 [(1867]). Cleve's newest investigations will give more detailed information on the "simplest of the platinum ammonium chloride compounds" [*Platinsalmiakverbindungen*],[38] which have long been known in any case. It is unfortunate that more accurate information is not given on which of the isomers were obtained by direct action of ammonia on platinous chloride.

After these general remarks, I proceed to a more detailed discussion of the *platinum bases*.

When in my above-cited work[2] it was necessary for me to prove my theoretical view of the metal bases, particularly in specific, more complicated cases, I chose my examples from Cleve's above-mentioned investigations, because they seemed to me to explain the most important, purely factual relationships in more detail and more completely than did any other investigations. Of special interest to me here were the many $\overset{\text{IV}}{\text{Pt}}$ compounds, which gave the most conspicuous proofs for the influence of the place which the Cl^-, OH^-, and acid atoms[39] occupy in the complex, as, for example, sulfatodiamminechloride, which is not precipitable with barium chloride:

$$\mathrm{Pt^2 \cdot 4\overline{NH}^3 \cdot O^2 \cdot \dddot{S}^2 \cdot \overline{Cl} \cdot \overline{H}}\ ^{40}$$

or, according to my view:

$$\mathrm{Pt}\begin{array}{l} \mathrm{O} \\ \mathrm{a \cdot a \cdot O} \\ \mathrm{a \cdot a \cdot Cl} \\ \mathrm{O \cdot H} \end{array}\!\!\begin{array}{l} \left.\begin{array}{l}\\ \\ \end{array}\right\} \mathrm{SO^2} \\ \\ \\ \end{array}\ ^{41}$$

[37] [Gros' chloride is *trans*-$[Pt(NH_3)_4Cl_2]Cl_2$ (J. Gros, *Ann.*, *27*, 241–256 (1838); *Ann. chim. phys.*, [2] *69*, 204–233 (1838)).—G.B.K.]

[38] [Platinum-ammines.—G.B.K.]

[39] [Anions.—G.B.K.]

[40] [Berzelius employed dots over atoms to indicate oxygen atoms. Thus $\dddot{S}^2 = S_2O_3$.—G.B.K.]

I also attempted to formulate atomistically the peculiar and certainly quite normally considered *pyridine compounds* of Anderson[42] (*Ann.*, *96*, 199[–205 (1855)]) which especially seemed to confirm my view. Thus I wrote the platosopyridine[43] and platinopyridine hydrochlorides[44] as

$$\overset{\mathrm{II}}{\mathrm{Pt}}\begin{matrix}\overset{\mathrm{III}}{\mathrm{N}}(\mathrm{C^5H^5})\cdot\mathrm{Cl}\\ \mathrm{N}(\mathrm{C^5H^5})\cdot\mathrm{Cl}\end{matrix} \quad \text{and} \quad \overset{\mathrm{IV}}{\mathrm{Pt}}\begin{matrix}\mathrm{Cl}\\ \mathrm{N}(\mathrm{C^5H^5})\cdot\mathrm{Cl}\\ \mathrm{N}(\mathrm{C^5H^5})\cdot\mathrm{Cl}\\ \mathrm{Cl}\end{matrix}$$

instead of

$$\mathrm{C^{10}\text{H̶}^4Pt\text{N̶},\ \text{H̶}Cl} \quad \text{and} \quad \mathrm{C^{10}\text{H̶}^3Pt\text{N̶},\ 2\text{H̶}Cl}.$$

Here a triatomic[3] radical is bound to unconditionally quinque-atomic[3] nitrogen.

For the present, I further cite some compounds that have long been known, that is, exactly the ones which Odling formulates as "ammonodiplatinum compounds" [*Ammondiplatinverbindungen*], namely, Hadow's Chloride,[45] Raewski's Nitrochloride,[46] and Gerhardt's Nitrate,[47] or according to Odling:

(1) $\mathrm{Cl^2O\overset{IV}{Pt}{}^2(H^5N^2)^4\cdot 4HCl}$,
(2) $\mathrm{Cl^2OPt^2(H^5N^2)^4\cdot 4H(NO^3)\cdot Ag}$,
(3) $\mathrm{(NO^3)^2OPt^2(H^5N^2)^4\cdot 4H(NO^3)\cdot Ag}$.

From Cleve's investigations we know that Hadow's Chloride also

---

41 [a = NH_3. This formula is given incorrectly as $\mathrm{Pt}\begin{matrix}\left.\begin{matrix}\mathrm{O}\\ \mathrm{a\cdot a}\end{matrix}\right\}\mathrm{SO^2}\\ \mathrm{a\cdot a\cdot Cl}\\ \mathrm{O\cdot H}\end{matrix}$.—G.B.K.]

42 [Thomas Anderson (Leith, Scotland, July 2, 1819—Chiswick, England, Nov. 2, 1874) investigated the nitrogenous bases in bone oil, discovering pyridine, lutidine, collidine, and picoline.—G.B.K.]

43 [*trans*-[$Pt(C_5H_5N)_2Cl_2$] (G. B. Kauffman, *Inorg. Syn.*, *7*, 249–253 (1963)).—G.B.K.]

44 [*cis*-[$Pt(C_5H_5N)_2Cl_4$], Anderson's salt.—G.B.K.]

45 [*trans*-[$Pt(NH_3)_4Cl(OH)]Cl_2$ (E. A. Hadow, *J. Chem. Soc.*, *19*, 345–352 (1866)).—G.B.K.]

46 [*trans*-[$Pt(NH_3)_4Cl(OH)](NO_3)_2$ (Raewsky, *Compt. rend.*, *23*, 253–254 (1846); *Ann. chim. phys.*, [3] *22*, 278–300 (1848)).—G.B.K.]

47 [*trans*-[$Pt(NH_3)_4(NO_3)OH](NO_3)_2$ (C. Gerhardt, *Compt. rend. trav. chim.*, *6*, 273–301 (1850)).—G.B.K.]

contains Ag. We thus write without hesitation (a = NH^3):

$$\begin{array}{ccc} (1) & (2) & (3) \\ \overset{IV}{Pt}\begin{array}{l} Cl \\ a\cdot a\cdot Cl \\ a\cdot a\cdot Cl \\ O\cdot H \end{array} & Pt\begin{array}{l} Cl \\ a\cdot a\cdot O\cdot NO^2 \\ a\cdot a\cdot O\cdot NO^2 \\ O\cdot H \end{array} & Pt\begin{array}{l} O\cdot NO^2 \\ a\cdot a\cdot O\cdot NO^2 \\ a\cdot a\cdot O\cdot NO^2 \\ O\cdot H \end{array} \end{array}$$

According to Cleve, Raewski's Salt,[46] for example, yields with HCl ordinary Gros' Chloride.[37] Cleve acquainted us with the true diplatinum compounds later (see the Correspondence).[23]

Finally, I should like to cite, atomistically formulated, as very enlightening examples of my view, the *double salts* of Buckton,[48] for example, $Pt\not{H}^6\not{N}^2\not{Cl} + Cu\not{Cl}$ (*Jahresbericht*, *1852*, 426) and of Thomsen[49] [*Jahresbericht*], *1868*, 278).

Thomsen gives the formulas:

$$Cu\not{Cl}, \not{N}\not{H}^3\not{H}Cl, \not{N}\not{H}^2Pt^{50} \quad \text{and} \quad Pt\not{Cl}, \not{N}\not{H}^3\not{H}Cl, \not{N}\not{H}^2Cu.^{51}$$

According to my method, they would be written:

$$Pt\begin{array}{l} a\cdot a\cdot Cl\cdot Cl \\ a\cdot a\cdot Cl\cdot Cl \end{array}Cu \quad \text{and} \quad Cu\begin{array}{l} a\cdot a\cdot Cl\cdot Cl \\ a\cdot a\cdot Cl\cdot Cl \end{array}Pt.$$

Nitrogen copulated to nitrogen and chlorine to chlorine in both cases facilitate the cohesion of the entire molecule, but, as the formulas show, in reverse order. Both are quite normal halide double salts[52] of different metal-ammonia bases.

Finally, in regard to the *nomenclature* of all these bases, I should like briefly to repeat here my earlier suggestion.

According to the above view, the bases are formed without exception by inserting into metal chlorides, for example, 1, 2, or $3NH^3$, so to speak, between the metal and the halide. Thus the newly formed bases could most simply be called mono-, di-, or triammonias. For the sake of brevity, however, the name *ammine* [*Ammin*], already long in use here and there, would be more suitable instead of ammonia.

---

[48] [George Bowdler Buckton (Oakfield, England, May 24, 1818—Haslemere, England, Sept. 25, 1905).—G.B.K.]

[49] [Hans Peter Jørgen Thomsen (Copenhagen, Denmark, Feb. 16, 1826—Copenhagen, Feb. 13, 1908), the Danish thermochemist.—G.B.K.]

[50] [$[Pt(NH_3)_4][CuCl_4]$ (G. B. Buckton, *Ann.*, *84*, 270 (1852); *Quart. J. Chem. Soc.*, *5*, 213–222 (1853)). The formula is incorrectly given as "$Cu\not{Cl}, \not{N}\not{H}^3\not{H}Cl, \not{N}\not{H}^2Pt$."—G.B.K.]

[51] [$[Cu(NH_3)_4][PtCl_4]$ (J. Thomsen, *Oversigt K. Dansk. Selskabs Forhl.*, *1867*, 225.—G.B.K.]

[52] [Actually, complex salts.—G.B.K.]

The prefix *di-*, *tri-*, and so forth (mono- for NH^3 does not require special mention) would then describe not the entire number of the ammonia atoms[53] entering into the complex, but simply the number of the ammonias chained to one another. The rest follows automatically from the well-known saturation capacity of the metal, especially if actual proof is provided for what experience acquired so far seems to show unequivocally, namely, that the extraradical points of attack of the metals once and for all lose the capacity for binding the nitrogen of the ammonia.

The constantly doubled *m* in the last syllable of ammine would be of special use in order to prevent confusion with the organic amines and diamines (of amine NH^2), such as ethylamine, diethylenediamine,[54] and so forth.

Besides, there is of course no obstacle to the use of the purely empirical expressions NH^3, N^2H^6, and N^3H^9 instead of the completely worked-out arrangement formula NH^3, $NH^3 \cdot NH^3$, and $NH^3 \cdot NH^3 \cdot NH^3$ for the mono-, di-, and triammonia.

Thus, for example, we would have:

$Au\,NH^3 \cdot Cl$ *Aurosammine* Chloride[55,56]

$$Cu\begin{matrix} NH^3 \cdot NH^3 \cdot Cl \\ NH^3 \cdot NH^3 \cdot Cl \end{matrix} \quad \text{or} \quad Cu\begin{matrix} N^2H^6 \cdot Cl \\ N^2H^6 \cdot Cl \end{matrix}$$ *Cupridiammine* Chloride[57]

$$Co\begin{matrix} NH^3 \cdot NH^3 \cdot NH^3 \cdot Cl \\ NH^3 \cdot NH^3 \cdot NH^3 \cdot Cl \end{matrix} \quad \text{or} \quad Co\begin{matrix} N^3H^9 \cdot Cl \\ N^3H^9 \cdot Cl \end{matrix}$$ *Cobaltotriammine* Chloride[58]

With special consideration to Cleve's investigations we wish somewhat more completely to cite, as extremely instructive examples, the entire series of platinum bases (as chlorides).

(A) Platoso Compounds $\overset{II}{Pt}$

(1) *Platodiammine*: $\overset{II}{Pt}\begin{matrix} NH^3 \cdot NH^3 \cdot Cl \\ NH^3 \cdot NH^3 \cdot Cl \end{matrix}$, *Reiset's* First Base,[19]

(2) *Platosammine*: $\overset{II}{Pt}\begin{matrix} NH^3 \cdot Cl \\ NH^3 \cdot Cl \end{matrix}$, *Reiset's* Second Base,[59]

---

[53] [Ammonia molecules.—G.B.K.]

[54] [Werner (*Z. anorg. Chem.*, *14*, 21–27 (1897)) later preserved this nomenclature with appropriate modifications.—G.B.K.]

[55] N.B. The chloride forms shiny, silver-white flakes.

[56] [$[Au(NH_3)]Cl$.—G.B.K.]

[57] [$[Cu(NH_3)_4]Cl_2$.—G.B.K.]

[58] [$[Co(NH_3)_6]Cl_2$ (G. B. Kauffman and N. Sugisaka, *Inorg. Syn.*, *9*, 157–160 (1967)).—G.B.K.]

[59] [*trans*-$[Pt(NH_3)_2Cl_2]$ (G. B. Kauffman and D. O. Cowan, *Inorg. Syn.*, *7*, 239–245 (1963)).—G.B.K.]

(3) *Platosemidiammine*: $\overset{\text{II}}{\text{Pt}}\begin{array}{l}\text{Cl}\\ \text{NH}^3\cdot\text{NH}^3\cdot\text{Cl}\end{array}$ [60,61]

(B) Platinicum Compounds $\overset{\text{IV}}{\text{Pt}}$ [62]

(4) *Platindiammine*, *Gros'* Base, [37]
(5) *Platinammine*, *Gerhardt's* Base,[63]
(6) *Platinsemidiammine*:[64]

$$\overset{\text{IV}}{\text{Pt}}\begin{array}{l}\text{Cl}\\ \text{NH}^3\cdot\text{NH}^3\cdot\text{Cl}\\ \text{NH}^3\cdot\text{NH}^3\cdot\text{Cl}\\ \text{Cl}\end{array}\qquad \overset{\text{IV}}{\text{Pt}}\begin{array}{l}\text{Cl}\\ \text{NH}^3\cdot\text{Cl}\\ \text{NH}^3\cdot\text{Cl}\\ \text{Cl}\end{array}\qquad \overset{\text{IV}}{\text{Pt}}\begin{array}{l}\text{Cl}\\ \text{Cl}\\ \text{NH}^3\cdot\text{NH}^3\cdot\text{Cl}\\ \text{Cl}\end{array}$$

(C) Diplatinum Compounds $\overset{\text{VI}}{\text{Pt}}$

(7) *Diplatindiammine*:

$$\begin{array}{ll} & \text{Cl}\\ \overset{\text{IV}}{\text{Pt}} & \begin{array}{l}\text{NH}^3\cdot\text{NH}^3\cdot\text{Cl}\\ \text{NH}^3\cdot\text{NH}^3\cdot\text{Cl}\end{array}\\ \overset{\text{IV}}{\text{Pt}} & \begin{array}{l}\text{NH}^3\cdot\text{NH}^3\cdot\text{Cl}\\ \text{NH}^3\cdot\text{NH}^3\cdot\text{Cl}\end{array}\\ & \text{Cl}\end{array}$$

---

[60] Of great interest is the seemingly complete agreement in the relationship of platinum to phosphorus even though, of course, the only really basic PR^3 must be used here. Of [A.] Cahours and [H.] Gal's compounds (*Compt. rend.*, *70*, 897–903 [1870]), the water-soluble one is no doubt $\text{Pt}\begin{array}{l}\text{PR}^3\cdot\text{PR}^3\cdot\text{Cl}\\ \text{PR}^3\cdot\text{PR}^3\cdot\text{Cl}\end{array}$ $[[Pt(R_3P)_4]Cl_2]$, the insoluble white one formed from this is $\text{Pt}\begin{array}{l}\text{PR}^3\cdot\text{Cl}\\ \text{PR}^3\cdot\text{Cl}\end{array}$ [*cis*-$[Pt(R_3P)_2Cl_2]$], and the directly obtained, insoluble yellow one is probably $\text{Pt}\begin{array}{l}\text{PR}^3\cdot\text{PR}^3\cdot\text{Cl}\\ \text{Cl}\end{array}$ [*trans*-$[Pt(R_3P)_2Cl_2]$]. The preference for forming diammines (or diphosphammines) appears very clearly here too.

[61] [*cis*-$[Pt(NH_3)_2Cl_2]$ (G. B. Kauffman and D. O. Cowan, *Inorg. Syn.*, *7*, 239–245 (1963)). The original incorrectly reads "*Platosamidiammine*."—G.B.K.]

[62] [The original incorrectly reads "$\overset{\text{II}}{\text{Pt}}$."—G.B.K.]

[63] [*trans*-$[Pt(NH_3)_2Cl_4]$ (G. B. Kauffman, *Inorg. Syn.*, *7*, 236–238 (1963)).—G.B.K.]

[64] [*cis*-$[Pt(NH_3)_2Cl_4]$ (G. B. Kauffman, *Inorg. Syn.*, *7*, 236–238 (1963)).—G.B.K.]

When I became familiar with the results of Cleve's investigations, it occurred to me that, if this view were correct, there still might be three (or four) platinum bases to be prepared, namely, *platosemiammine*, *platinsemiammine*, and *diplatinammine* (perhaps also of a *monodiammine*):

$$\overset{\text{II}}{\text{Pt}}\begin{matrix}\text{Cl}\\ \text{a}\cdot\text{Cl},\end{matrix}\qquad \overset{\text{IV}}{\text{Pt}}\begin{matrix}\text{Cl}\\ \text{Cl}\\ \text{a}\cdot\text{Cl},\\ \text{Cl}\end{matrix}\quad \text{and}\quad \overset{\text{VI}}{\text{Pt}}\begin{matrix}\text{Cl}\\ \text{a}\cdot\text{Cl}\\ \text{a}\cdot\text{Cl}\\ \text{a}\cdot\text{Cl}\\ \text{a}\cdot\text{Cl}\\ \text{Cl}\end{matrix}\qquad \overset{\text{VI}}{\text{Pt}}\begin{matrix}\text{Cl}\\ \text{a}\cdot\text{Cl}\\ \text{a}\cdot\text{a}\cdot\text{Cl}\\ \text{a}\cdot\text{a}\cdot\text{Cl}\\ \text{a}\cdot\text{Cl}\\ \text{Cl}\end{matrix}$$

Yet in any case the formulas for the semiammines, so called here for the sake of brevity, are in part not yet conclusively determined. Cleve's attempts to prepare the last-named bases seem to indicate that perhaps formula (3) should be doubled. Compounds of a tetraatomic diplatinum (or diplatosam) $\overset{\text{IV}}{\not{\text{Pt}}}$ ($\overset{\text{IV}}{\text{Pt}} = = \overset{\text{IV}}{\text{Pt}}$) as well as the existence of still more complicated compounds derived from them (for example, of $\overset{\text{IV}}{\text{Pt}}$) would also be not at all theoretically improbable. The striking similarity of the combining laws of platinum to those of carbon becomes clearer and clearer.[65]

In regard to the *cobalt bases*, on which I hope soon to be able to report more extensively, for the present I must confine myself to a few short remarks, because too much space has already been taken up.

In order to confirm my general view of metal bases, these substances were from the beginning especially important to me because of their complicated composition.

I believed ([*Die*] *Chemie der Jetztzeit*, p. 294) that luteocobaltammonia should immediately be formed from cobalt-triammonias by the action of iodine because of the very easy formation of the dicobaltin, according to the equation:[66]

[65] [According to Lev Aleksandrovich Chugaev, "Each element of the platinum group represents, as it were, its own miniature organic chemistry, in which, as if in a mirror, are reflected the features of its chemical individuality" (G. B. Kauffman, *J. Chem. Educ.*, *40*, 656–665 (1963)).—G.B.K.]

[66] [During the first half of the nineteenth century, measurement of vapor density was the only method for determining molecular weights. Until the classic work of Raoult and van't Hoff on colligative properties of solutions (1882 *et seq.*) no reliable method existed for determining molecular weights of nonvolatile compounds. Thus, cobalt(III) chloride was thought to have the composition

$$
\begin{matrix}
\overset{\text{II}}{\text{Co}}\begin{matrix}\text{a}\cdot\text{a}\cdot\text{a}\cdot\text{O}\\ \text{a}\cdot\text{a}\cdot\text{a}\cdot\text{O}\end{matrix}\text{SO}^2 \\[1em]
\overset{\text{II}}{\text{Co}}\begin{matrix}\text{a}\cdot\text{a}\cdot\text{a}\cdot\text{O}\\ \text{a}\cdot\text{a}\cdot\text{a}\cdot\text{O}\end{matrix}\text{SO}^2
\end{matrix}
+ \text{J}^2 =
\left\{
\begin{matrix}
\text{J} \\
\overset{\text{IV}}{\text{Co}}\begin{matrix}\text{a}\cdot\text{a}\cdot\text{a}\cdot\text{O}\\ \text{a}\cdot\text{a}\cdot\text{a}\cdot\text{O}\end{matrix}\text{SO}^2 \\[1em]
\overset{\text{IV}}{\text{Co}}\begin{matrix}\text{a}\cdot\text{a}\cdot\text{a}\cdot\text{O}\\ \text{a}\cdot\text{a}\cdot\text{a}\cdot\text{O}\end{matrix}\text{SO}^2 \\
\text{J}
\end{matrix}
\right.
$$

This method should unquestionably be the simplest and most productive of all those proposed so far for the preparation of cobaltin bases. In a half hour, any desired amounts of the salt can be prepared, even though it is most advantageous to work with smaller portions. The ammoniacal solution of the cobalt sulfate is heated for some time with iodine (Co:J).[67] The yellow *dicobaltitriammine iodosulfate* is separated off without difficulty as a very difficultly soluble substance from the simultaneously formed red *dicobaltinditriammine* (roseo) *iodosulfate*, which is very readily soluble in ammonia.

In regard to the different derivatives of these mixed salts, for the present I note only the great stability of the, of course, remarkably normal luteo or, as I am now usually inclined to assume, pure triammine (instead of monoditriammine) complex. If the iodosulfate is treated with chlorine, for example, the combined ammonia is not destroyed, neither is the iodine simply split off, but, as in the presence of true alkalies, it is in part continually bound in an oxidized state. The investigations in this laboratory will be continued.

Lund, Jan. 5, 1871.

Co_2Cl_6 (by analogy with volatile Fe_2Cl_6), and hence cobalt-ammines were considered dimers, as formulated here by Blomstrand. It was not until S. M. Jørgensen (*J. prakt. Chem.* [2], *41*, 429–439 (1890)) and J. Petersen (*Z. physik. Chem.*, *10*, 580–592 (1892)) deduced evidence for monomeric molecular weights by freezing-point and conductance measurements of such solutions that Blomstrand's original formulas were halved. Inasmuch as the concept of octahedral configuration based on coordination number six was a fundamental postulate of Werner's theory from its inception, it is possible that without Jørgensen's halving of Blomstrand's formulas (e.g., $Co_2Cl_6 \cdot 12NH_3$ became $CoCl_3 \cdot 6NH_3$), the coordination theory might never have been conceived. The equation given by Blomstrand should read: $[Co(NH_3)_6]SO_4 + \frac{1}{2}I_2 \rightarrow [Co(NH_3)_6](SO_4)I$.—G.B.K.]

[67] [This should read Co:I = 1 (J. M. Krok, *Acta Univ. Lund.*, *1870*, 15; *Ber.*, *4*, 711–717 (1871); C. W. Blomstrand, *J. prakt. Chem.*, [2] *3*, 186–224 (1871), p. 206).—G.B.K.]

Christian Wilhelm Blomstrand (1826–1897)
[Courtesy of Lunds Universitetsbibliotek, Lund, Sweden.]

SOPHUS MADS JØRGENSEN (1837–1914)

[From *Proceedings of the Symposium on Co-ordination Chemistry, Copenhagen, August 9–13, 1953*, p. 14. Copenhagen: Danish Chemical Society, 1954.]

INTRODUCTION TO PAPER SIX*

[In 1893 a little-known twenty-six-year-old *Privat-Dozent* at the Eidgenössisches Polytechnikum in Zürich arose to challenge and discard the confining rigidities of both the Kekulé valence theory and the Blomstrand–Jørgensen chain theory.[1] Like a modern scientific Alexander, Alfred Werner cut the Gordian knot that for decades had caused confusion in structural inorganic chemistry. The era of coordination chemistry had begun.

In his revolutionary theory,[2,3] which marked an abrupt break with the classical theories of valence and structure, Werner postulated two types of valence—primary or ionizable (*Hauptvalenz*) and secondary or non-ionizable (*Nebenvalenz*). According to the theory, every metal in a particular oxidation state (primary valence) has a definite coordination number, i.e., a fixed number of secondary valences that must be satisfied. Whereas primary valences can be satisfied only by anions, secondary valences can be satisfied not only by anions but also by neutral molecules such as ammonia, water, organic amines, sulfides, and phosphines. These secondary valences are directed in space around the central metal ion (octahedral for coordination number 6; square planar or tetrahedral for coordination number 4), and the combined aggregate forms a "complex," which should exist as a discrete unit in solution.

The acknowledged test of a scientific theory is its ability to explain known facts and to predict new ones. In examining the success of Werner's coordination theory in meeting these criteria, we shall consider two aspects of the metal-ammines—(1) constitution, i.e., the manner of bonding of the constituent atoms and groups, and (2) configuration, i.e., the spatial arrangement of these atoms and groups. Although we shall confine ourselves primarily to compounds of coordination number six (cobalt(III)), we should bear in mind that Werner used similar arguments to prove the constitution and configuration for compounds of coordination number four.

---

*A portion of this Introduction has been adapted from G. B. Kauffman, "Alfred Werner's Coordination Theory: A Brief Historical Introduction," *Educ. Chem.*, *4*, 11–18 (1967).

[1] G. B. Kauffman, (*a*) *Alfred Werner—Founder of Coordination Chemistry*, Springer-Verlag, Berlin, Heidelberg, New York, 1966; (*b*) *Chemistry*, *39* (12), 14–18 (1966); (*c*) *Educ. Chem.*, *4*, 11–18 (1967).

[2] G. B. Kauffman, *Classics in Coordination Chemistry, Part I: The Selected Papers of Alfred Werner*, Dover Publications, Inc., New York, 1968.

[3] A. Werner, *Z. anorg. Chem.*, *3*, 267–330 (1893). For a discussion and English translation see Kauffman, *Classics*, pp. 5–88.

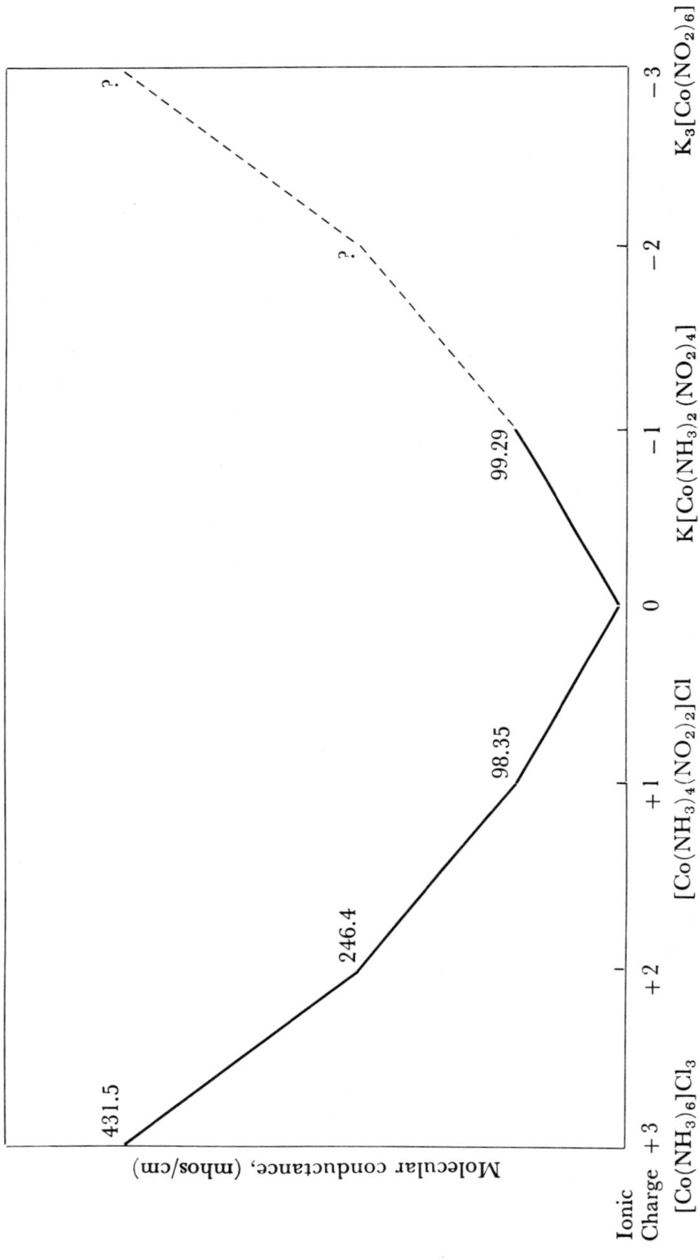

Fig. 1. Conductivities of cobalt(III) coordination compounds (Werner and Miolati, 1894).[1c]

Werner's first published experimental work in support of his coordination theory was a study of conductivities,[4] carried out during the years 1893 to 1896 in collaboration with his friend and former fellow student Arturo Miolati (1869–1956),[5] who was later to become Professor of Chemistry at the Universities of Turin and Padua. According to the new theory, the charge of a complex ion should be equal to the algebraic sum of the charges of the central metal ion and of the coordinated groups. Consequently, as neutral molecules of ammonia (A) in a metal-ammine (MA_6) are successively replaced by anions (B), the number of ions in the resulting compounds should progressively decrease until a nonelectrolyte is formed and then increase as the complex becomes anionic.

Friedrich Kohlrausch's principle of the additivity of equivalent conductivities of salts (1879)[6] provided Werner and Miolati with a convenient method for determining the number of ions in a variety of complexes. After having established the ranges of conductivities to be expected for salts of various types, they were able to demonstrate the complete agreement in magnitude, variation, and pattern between their experimentally measured conductivities (Fig. 1) and those predicted according to the coordination theory. Their results were also concordant with the number of precipitable halogen atoms. The constitutions and predicted numbers of ions according to the two theories are contrasted in Table 1.

For compounds of the first three classes, the electrolytic character as predicted by the two theories is in complete agreement, and conductivity data do not permit a choice between the two. For triammines, however, the ionic character differs radically according to the two theories, and the conductivities of these compounds became an important and bitterly contested issue. For some nonelectrolytes, unfortunately, Werner and Miolati's conductivity values were not always zero because of aquation reactions:

$$[Co(NH_3)_3Cl_3]^0 + H_2O \rightleftharpoons [Co(NH_3)_3(H_2O)Cl_2]^+ + Cl^-.$$

Jørgensen immediately seized upon such "discrepancies" in an attempt to discredit their results. But in its explanation of anionic complexes and its demonstration of the existence of a continuous transition series (*Übergangsreihe*) between metal-ammines (MA_6) and double salts (MB_6), the Werner theory succeeded in an area in which the Blomstrand–Jørgensen theory could not pretend to compete.

The technique of "isomer counting" as a means of proving configuration and its geometric consequences with respect to the number of isomers

[4] A. Werner and A. Miolati, *Z. physik. Chem.*, *12*, 35–55 (1893); *ibid.*, *14*, 506–521 (1894); *ibid.*, *21*, 225–238 (1896). For discussions and English translations of the first two papers see Kauffman, *Classics*, pp. 89–139.

[5] G. B. Kauffman, *Isis*, *61*, 251–253 (1970).

[6] F. Kohlrausch, *Ann. Phys.*, *11*, 653 (1880).

TABLE 1[1c]

Class of Compound	BLOMSTRAND–JØRGENSEN Formula	BLOMSTRAND–JØRGENSEN No. of Ions	WERNER Formula	WERNER No. of Ions
Hexaammines MA_6	$Co-NH_3-NO_2$ $Co-NH_3-NO_2$ $Co-NH_3-NH_3-NH_3-NH_3-NO_2$ ↓ $-NH_3$	4	$[Co(NH_3)_6](NO_2)_3$ ↓ $-NH_3$	4
Pentaammines MA_5B	$Co-NO_2$ $Co-NH_3-NO_2$ $Co-NH_3-NH_3-NH_3-NH_3-NO_2$ ↓ $-NH_3$	3	$[Co(NH_3)_5NO_2](NO_2)_2$ ↓ $-NH_3$	3
Tetraammines MA_4B_2	$Co-NO_2$ $Co-NO_2$ $Co-NH_3-NH_3-NH_3-NH_3-NO_2$ ↓ $-NH_3$	2	$[Co(NH_3)_4(NO_2)_2]NO_2$ ↓ $-NH_3$	2
Triammines MA_3B_3	$Co-NO_2$ $Co-NO_2$ $Co-NH_3-NH_3-NH_3-NO_2$	2	$[Co(NH_3)_3(NO_2)_3]$ ↓ $-NH_3$	0
Diammines MA_2B_4	Unaccountable	—	$K[Co(NH_3)_2(NO_2)_4]$ ↓ $-NH_3$	2
Monoammines MAB_5	Unaccountable	—	Unknown for Cobalt ↓ $-NH_3$	(3)
Double Salts, MB_6	Unaccountable	—	$K_3[Co(NO_2)_6]$	4

expected had been considered as early as 1875 by Jacobus Henricus van't Hoff (1852–1911),[7] and the general method is probably most familiar through Wilhelm Körner's work of 1874 on disubstituted and trisubstituted benzene derivatives.[8] Yet the technique of comparing the number and type of isomers actually prepared with the number and type theoretically predicted for various configurations probably reached the height of its development with Werner's work. By this method, he was able not only to discredit completely the rival Blomstrand–Jørgensen chain theory but also to demonstrate unequivocally that trivalent cobalt possesses an octahedral configuration rather than another possible symmetrical arrangement such as hexagonal pyramidal, hexagonal planar, or trigonal prismatic. The method is summarized in Fig. 2 and Table 2.

As a comparison of Columns *IV* and *V* of Table 2 shows, in most cases, the number and type of isomers prepared corresponded to the expecta-

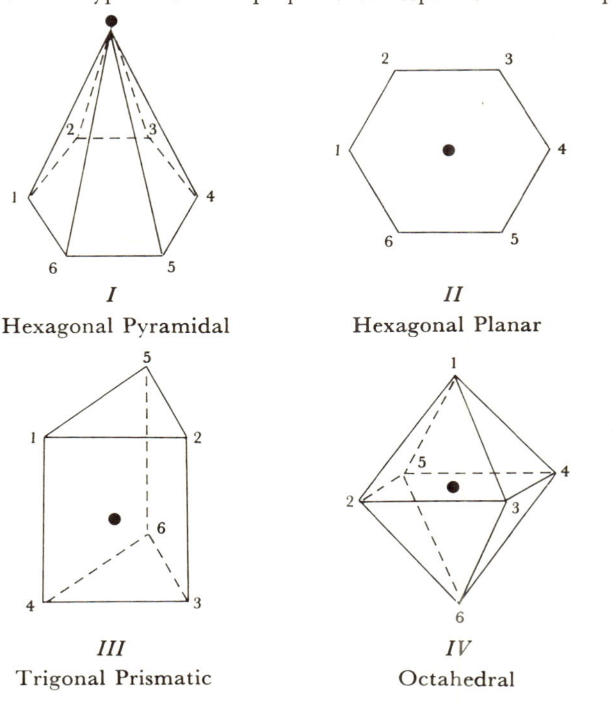

Fig. 2. Configurational possibilities for coordination number six

[7] J. H. van't Hoff, *Maandblad voor Natuurwetenschappen*, *6*, 37 (1875).
[8] W. Körner, *Gazz. chim. ital.*, *4*, 305–446 (1874); *Über die Bestimmung des chemischen Ortes bei den aromatischen Substanzen*, Ostwald's Klassiker der exakten Wissenschaften No. 174, Wilhelm Engelmann, Leipzig, 1910.

Table 2 Proof of Configuration for Coordination Number Six by "Isomer Counting"

Compound type	Theoretically predicted isomers: I Hexagonal pyramidal	II Hexagonal planar (a special case of *I*)	III Trigonal prismatic†	IV Octahedral	Experimentally found isomers: V	Result VI
MA_6	One form only	One form only	One form only	One form only	One form only	None
MA_5B	One form only	One form only	One form only	One form only	One form only	None
$M(\overline{AA})B_4$*	One form only	One form only	Two geometric	One form only	One form only	Provisionally eliminates trigonal prismatic (*III*)
MA_4B_2	Three geometric (1,2; 1,3; 1,4)	Three geometric (1,2; 1,3; 1,4)	Three geometric (1,2; 1,3; 1,4)	Two geometric (1,2 *cis*; 1,6 *trans*)	Two or less geometric	Provisionally proves octahedral (*IV*); discovered 1907
MA_3B_3	Three geometric (1,2,3; 1,2,4; 1,3,5)	Three geometric (1,2,3; 1,2,4; 1,3,5)	Three geometric (1,2,3; 1,2,5; 1,2,6)	Two geometric (1,2,3 *facial*; 1,2,6 *peripheral*)	Two or less geometric	Provisionally proves octahedral (*IV*)
$M(\overline{AA})_2B_2$* *or* $M(\overline{AA})_2BC$*	Two *geometric*	Two *geometric*	Four *geometric*, one of which is asymmetric	Two *geometric* (1,2 *cis*; 1,6 *trans*), the first of which is *asymmetric*	Two *geometric* (1.2 *cis*; 1,6 *trans*), the first of which was *resolved*	Unequivocally proves octahedral (*IV*); discovered 1911
$M(\overline{AA})_3$*	One form only	One form only	Two *geometric*	One asymmetric pair	One pair *optical* resolved	Unequivocally proves octahedral (*IV*); discovered 1912

* $\overline{AA}$ represents a symmetrical bidentate (chelate) ligand. Such ligands coordinate at two adjacent positions. They can span *cis* positions but not *trans* positions.

† Coordination compounds with this configuration have recently been synthesized (R. Eisenberg and J. A. Ibers, *J. Am. Chem. Soc.*, *87*, 3776–3778 (1965); *Inorg. Chem.*, *5*, 411–416 (1966); E. I. Stiefel and H. B. Gray, *J. Am. Chem. Soc.*, *87*, 4012–4013 (1965); H. B. Gray, R. Eisenberg, and E. I. Stiefel, in *Werner Centennial*, G. B. Kauffman, Symposium Chairman, American Chemical Society, Washington, D.C., 1967, pp. 641–650).

tions for the octahedral arrangement, but there were a few exceptions, and Werner required more than twenty years to accumulate a definitive proof for his structural ideas. For example, the best known case of geometrical (*cis-trans*) isomerism was observed (by Jørgensen)[9] not among simple tetraammines MA_4B_2 but among salts $M(\overline{AA})_2B_2$, in which the four ammonia molecules had been replaced by two molecules of the bidentate (chelate) organic base, ethylenediamine (en), i.e., among the so-called praseo (green) and violeo (violet) series of formula $CoCl_3 \cdot 2en$. Jørgensen regarded the difference in color as due to *structural* isomerism connected with the linking of the two ethylenediamine molecules, whereas Werner regarded the compounds as *stereoisomers*, i.e., compounds composed of the same atoms and bonds but differing only in the orientation of these atoms and bonds in space:

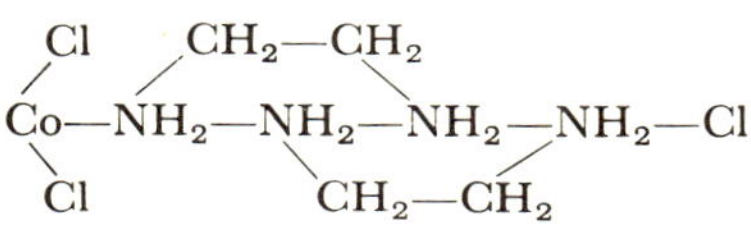

Praseo

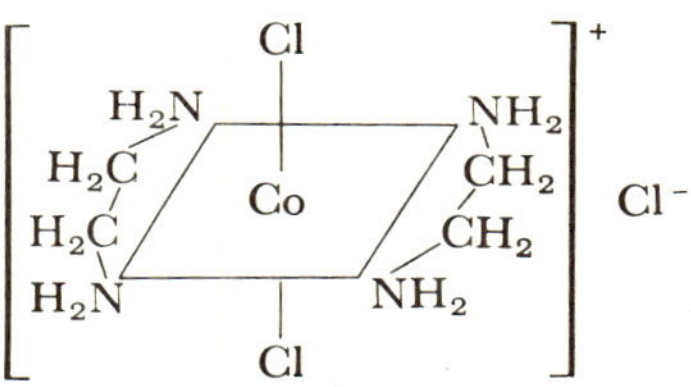

Jørgensen

Werner

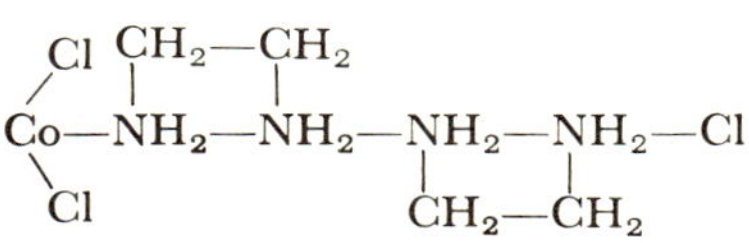

Violeo

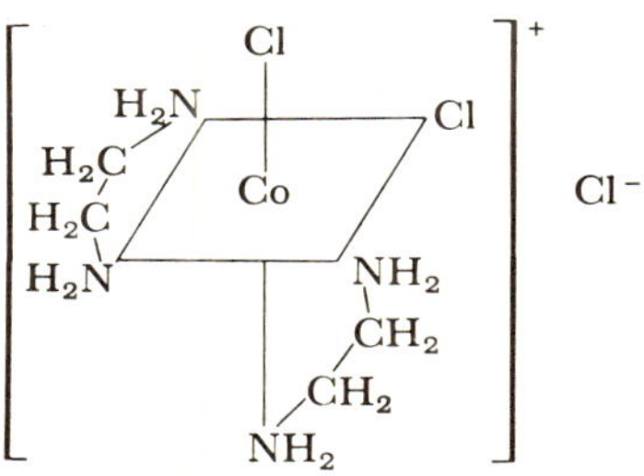

Jørgensen's and Werner's formulae for Praseo and Violeo ethylenediamine isomers.[1c]

If this type of isomerism were merely a geometrical consequence of the octahedral structure as Werner maintained, it should also be observed among simple tetraammines MA_4B_2 which do not contain ethylenediamine. Yet for compounds $[Co(NH_3)_4Cl_2]X$, only one series (praseo, green) was known.[10] Jørgensen, a confirmed empiricist, quite correctly criticized Werner's theory on the ground that it implied the existence of unknown compounds. It was not until 1907 that Werner finally

---

[9] S. M. Jørgensen, *J. prakt. Chem.*, [2] *39*, 1–26 (1889).

[10] W. Gibbs and F. A. Genth, *Am. J. Sci.*, [2] *23*, 248 (1857); *Ann. Chem.*, *104*, 155 (1857).

succeeded in synthesizing the unstable, highly crucial violeo tetraammines, *cis*-$[Co(NH_3)_4Cl_2]X$,[11]

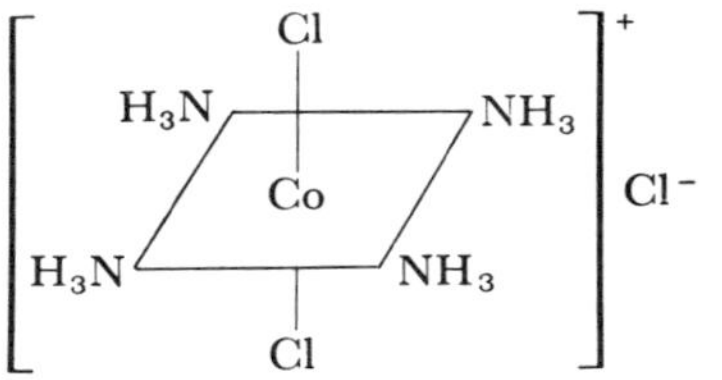

Praseo (*trans:* 1,6)
Gibbs and Genth (1857)

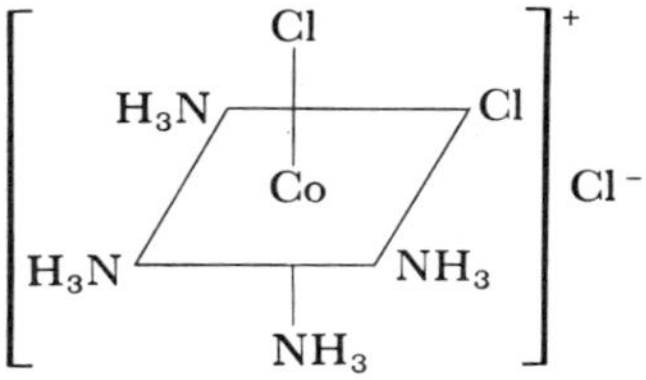

Violeo (*cis:* 1,2)
Werner (1907)

which were a necessary consequence of his theory but not of Jørgensen's. His Danish opponent immediately conceded defeat.

Even though the discovery of the long-sought violeo salts convinced Jørgensen that his own views could not be correct, Werner's success in preparing two, and only two, isomers for compounds of types MA_4B_2 and MA_3B_3 was not sufficient to prove conclusively his octahedral configuration. Despite such "negative" evidence, it could still be argued logically that failure to isolate a third isomer did not necessarily prove its nonexistence. A more "positive" type of proof was necessary.

As early as 1899,[12] Werner recognized that the resolution into optical isomers of certain types of coordination compounds containing chelate groups, which can span *cis* positions only, could provide the "positive" proof that he needed (Table 2). After many unsuccessful attempts, in 1911 he succeeded. His resolution, with his American student Victor King (1886–1958), of *cis*-chloroamminebis(ethylenediamine)cobalt(III) salts by means of the resolving agent silver *d*-α-bromocamphor-π-sulfonate was sufficient to prove conclusively the octahedral configuration for cobalt(III)[13–15]:

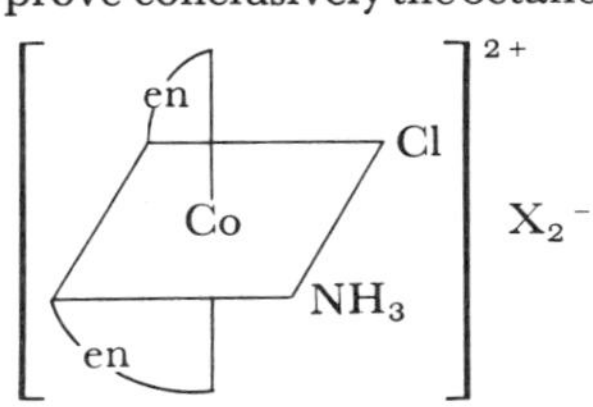

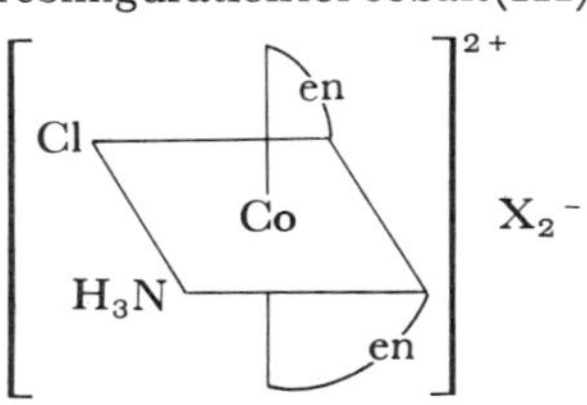

Yet because of the prevalent view that optical activity was almost always connected with carbon atoms, a number of Werner's contemporaries argued that the optical activity of these and the many other mononuclear

[11] A. Werner, *Ber.*, *40*, 4817–4825 (1907). For a discussion and English translation see Kauffman, *Classics*, pp. 141–154.
[12] A. Werner and A. Vilmos, *Z. anorg. Chem.*, *21*, 145–158 (1899).
[13] A. Werner, *Ber.*, *44*, 1887–1898 (1911). For a discussion and English translation see Kauffman, *Classics*, pp. 155–173.
[14] V. L. King, *J. Chem. Educ.*, *19*, 345 (1942).
[15] G. B. Kauffman, *Isis*, *66*, 38–62 (1975).

and polynuclear coordination compounds subsequently resolved by him was somehow due to the organic chelate groups present, even though these symmetrical ligands were all optically inactive. Any vestige of doubt was finally dispelled by Werner's resolution[16] in 1914 of completely carbon-free coordination compounds—the tris[tetraammine-μ-dihydroxocobalt(III)]cobalt(III) salts,

$$\left[\mathrm{Co}\left\{\begin{matrix}\mathrm{HO} & \\ & \mathrm{Co(NH_3)_4} \\ \mathrm{HO} & \end{matrix}\right\}_3\right]\mathrm{X_6}.$$

These salts are compounds of the $\mathrm{M}(\overline{\mathrm{AA}})_3$ type, in which $\overline{\mathrm{AA}}$ is the inorganic bidentate ligand

$$\left[\begin{matrix}\mathrm{HO} & \\ & \mathrm{Co(NH_3)_4} \\ \mathrm{HO} & \end{matrix}\right]^+$$

Sophus Mads Jørgensen,[17] Alfred Werner's primary scientific adversary, was born in Slagelse, Seeland, Denmark, Zeise's birthplace, on July 4, 1837. In 1857 he entered the University of Copenhagen from which he received his master's degree in chemistry in 1863. The following year he became assistant to Prof. Edward Augustus Scharling at the university. In 1867 he was appointed director of the chemical laboratories and, like Zeise before him, instructor at the Polytechnic School. His dissertation *Polyiodides of Alkaloids* earned him the doctorate in 1869. Two years later he was appointed lecturer at the university and in 1887, Professor of Chemistry, a position which he held until 1908, the year of his retirement. He died in Copenhagen on April 1, 1914.

Except for some early isolated research, Jørgensen devoted himself exclusively to investigating the coordination compounds of cobalt,

---

[16] A. Werner, *Ber.*, *47*, 3087–3094 (1914). For a discussion and English translation see Kauffman, *Classics*, pp. 175–184.

[17] S. P. L. Sørensen, *Fysisk Tidsskrift*, *12*, 217 (1913–1914); *Oversigt over Videnskabernes Selskabs Forhandlinger*, *1914*, 46–49; A. Werner, *Chemiker-Zeitung*, *38*, 557–564 (1914); S. Veibel, *Dansk Biografisk Leksikon*, Schultz Forlag, Copenhagen, 1937, Vol. 12, pp. 253–256; N. Bjerrum, *Proceedings of the Symposium on Co-ordination Chemistry, Copenhagen, August 9–13, 1953*, Danish Chemical Society, Copenhagen, 1954, pp. 12–16; G. B. Kauffman, *J. Chem. Educ.*, *36*, 521–527 (1959); reprinted in *Selected Readings in the History of Chemistry*, compiled by A. J. Ihde and W. F. Kieffer, Journal of Chemical Education, Easton, Penna., 1965, pp. 185–191; *idem*, *Chymia*, *6*, 180–204 (1960); *idem*, *Dictionary of Scientific Biography*, ed. by C. C. Gillispie, Charles Scribner's Sons, New York, 1973, Vol. 7, pp. 179–180; P. S. Cohen, "Effect of the Fixity of Ideas on the Werner-Jørgensen Controversy," in *Werner Centennial*, G. B. Kauffman, Symposium Chairman, Advances in Chemistry Series, American Chemical Society, Washington, 1967, pp. 8–40. For a list of Jørgensen's publications see S. Veibel, *Kemien i Danmark*, Nyt Nordisk Forlag, Arnold Busck, Copenhagen, 1943, Vol. II, pp. 253–256.

chromium, rhodium, and platinum, and this work, upon which his fame securely rests, forms an interconnected and continuous chain from 1878 to 1906. Although Werner's ideas eventually triumphed, Jørgensen's experimental observations are thereby in no way invalidated. On the contrary, his experiments, performed with meticulous care, have proved completely reliable and have provided the experimental foundation not only for the Blomstrand–Jørgensen theory but for Werner's coordination theory as well.

As a research worker, Jørgensen was methodical, deliberate, painstaking, and solitary. Although he could have delegated much routine work to assistants, he insisted on personally performing all his analyses, a task for which he reserved one day a week. In terms of Wilhelm Ostwald's twofold classification of scientific genius (classic *vs.* romantic),[18] Jørgensen would seem to embody the classic type—the slow and deep-digging completer who produces with long deliberation and slowly develops a traditional theory to new consequences. Considering his passion for perfection, his research output was tremendous, and we are indebted to him for many of the basic experimental facts of coordination chemistry.

The controversy between Jørgensen and Werner constitutes an excellent example of the synergism so often encountered in the history of science. Jørgensen's strong and conservative sense of history caused him to regard Werner's new theory as an unwarranted break in the development of theories of chemical structure. He regarded it as an *ad hoc* explanation insufficiently supported by experimental evidence. In their scholarly rivalry each chemist did his utmost to prove his views, and in the process a tremendous amount of fine experimental work was performed by both. Not all of Jørgensen's criticisms were valid, but in a number of cases Werner was forced to modify various aspects of his theory.

Although Jørgensen created no new structural theory of his own, he logically and consistently extended and modified Blomstrand's chain theory in order to interpret the many new series of complexes that he, Jørgensen, had succeeded in preparing for the first time. Ironic as it may seem, Jørgensen's work bore the seeds of the Blomstrand–Jørgensen theory's destruction, for many of the compounds first prepared by Jørgensen later proved instrumental in demonstrating the validity of Werner's views. The following paper (Paper Six),[19] which summarizes Jørgensen's more than twenty years of work on complexes, contains his final comparison between the two theories. In it Jørgensen insists that Blomstrand's theory was the only one compatible with the experimental facts. This, his final plea on behalf of Blomstrand's views, marks his withdrawal from the controversy.—G.B.K.]

---

[18] W. Ostwald, in *Das Buch der grossen Chemiker*, ed. by G. Bugge, Verlag Chemie, Berlin, 1929, Vol. 1, p. 405; E. Farber, *J. Chem. Educ.*, *30*, 600–604 (1953).
[19] S. M. Jørgensen, *Z. anorg. Chem.*, *19*, 109–157 (1899).

6

Sophus Mads Jørgensen

The Constitution of Cobalt, Chromium and Rhodium Bases. Part XI.

[Translation of "Zur Konstitution der Kobalt-, Chrom- und Rhodiumbasen. XI. Mitteilung," *Zeitschrift für Anorganische Chemie, 19,* 109–157 (1899).]

After Werner, in his article on triammine- and diamminecobalt salts,[1] abandoned those types of isomerism which were based on the assumption that the electronegative ions lay in different principal parts of his octahedron and after I abandoned the assumption that water could not occur instead of ammonia in my four-membered ammonia group,[2] just as I do not presently have any compelling reason to assume unlike valences in the trivalent cobalt atom,[3] after all this there is no longer any difficulty in comparing the two theories and in forming an opinion about the two conceptions. Although the change is quite insignificant, the comparison is made easier by the fact that Werner is now satisfied with a square and an axis perpendicular to it instead of an octahedron. The 4 atoms or atomic groupings (NH_3, OH_2), which Werner imagines as lying in the 4 corners of the square, thus correspond exactly to my four-membered ammonia group, in which now, according to results to date, one or two NH_3 molecules can be replaced by OH_2. Those atoms or groups of atoms, which Werner imagines as lying at the ends of the perpendicular axis, correspond to those which I assume to be bound to the metal atom. Werner's formulas can therefore be translated directly into mine. To take a few examples from Werner's most

[1] *Z. anorg. Chem., 15,* 143[–172 (1897)].
[2] *Ibid., 11,* 448 ff. [416–453 (1896)] and *14,* 417–418 [404–422 (1897)].
[3] *Ibid., 14,* 410 [404–422 (1897)].

recent investigations,[4] the isomerism (which, incidentally, is still uncertain) which Werner formulates as (a = NH_3, h = OH_2):

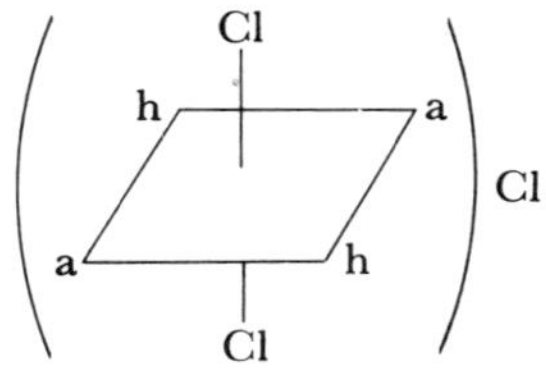

and

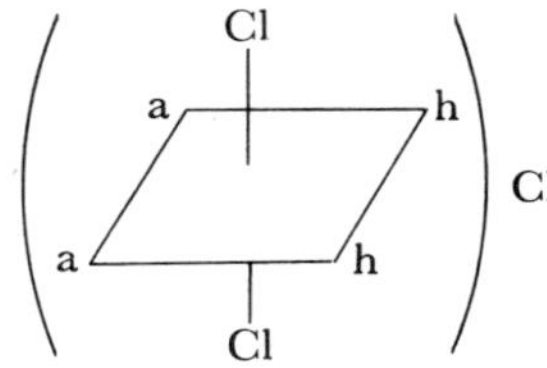

according to my theory becomes

$$\begin{matrix} \cdot Cl \\ Co\cdot a\cdot h\cdot a\cdot h\cdot Cl \\ \cdot Cl \end{matrix} \quad \text{and} \quad \begin{matrix} \cdot Cl \\ Co\cdot a_2\cdot h_2\cdot Cl \\ \cdot Cl \end{matrix}$$

Likewise the following formulas of Werner's:

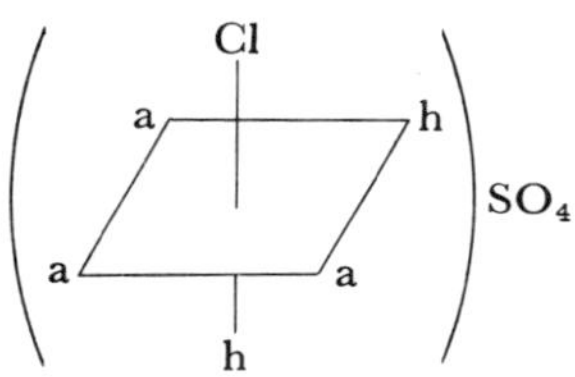

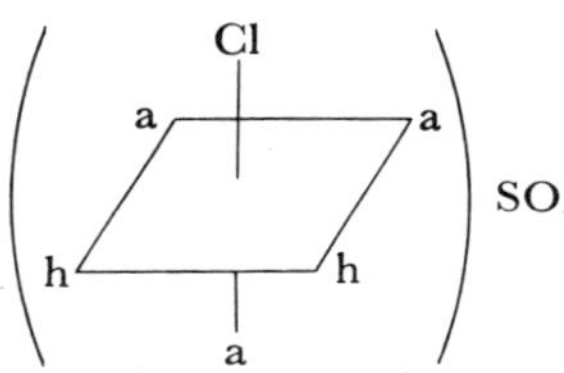

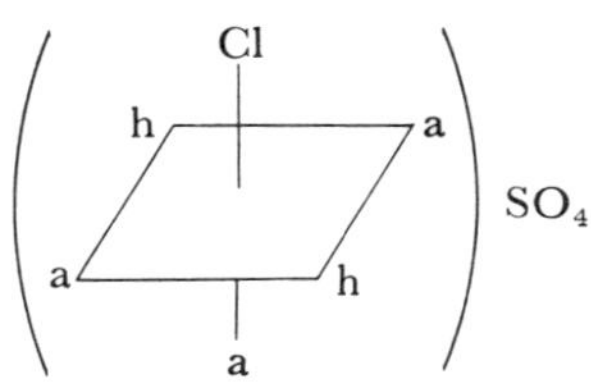

can be translated simply into my theory as:

$$\begin{matrix} \cdot Cl \\ Co\cdot a\cdot a\cdot a\cdot h\cdot \\ \cdot OH_2 \text{—}\cdot \end{matrix} SO_4 \qquad \begin{matrix} \cdot Cl \\ Co\cdot a\cdot a\cdot h\cdot h\cdot \\ \cdot NH_3 \text{—}\cdot \end{matrix} SO_4 \qquad \begin{matrix} \cdot Cl \\ Co\cdot a\cdot h\cdot a\cdot h\cdot \\ \cdot NH_3 \text{—}\cdot \end{matrix} SO_4$$

Such compounds will not fit into my theory only if Werner introduces an electronegative radical into one of the corners of his square (that is, as a non-ion). Such cases are extremely rare, and in none of these can Werner prove his formulation. They are the following:

1. The ammonia-violeo salts isomeric with the ammonia-praseo salts. Recently[5] I have explained so thoroughly that there is not a

[4] *Z. anorg. Chem.*, *15*, 149 and 152 [143–172 (1897)].
[5] *Z. anorg. Chem.*, *14*, 406 ff. [404–422 (1897)].

single reason for assuming the existence of ammonia-violeo salts that it seems unnecessary to refer again to it here. Everything indicates that this isomerism occurs only among the ethylenediaminecobalt salts and similar compounds.[6] This isomerism is not made more probable by the fact that Werner now[7] classes the chloroaquotetrammine salts with the violeo salts whereas he formerly[8] regarded them as belonging to the praseo series.[9] The only isomerism which is actually known and upon which Werner's assumption of ammonia-violeo salts is exclusively based is that between croceo and flavo salts.[10] But first of all these are nitrite salts, whose isomerism can be understood quite easily, among other things, by the difference between nitrites and nitro compounds.[11] In this case, however, the essential difference between the two series of salts, viz., their behavior toward acids, is of a type such that he points out explicitly that the isomerism lies in a difference between the nitro groups, while nothing whatsoever indicates that the ammonia groups of the two series are bonded differently.

2. While Werner's formula for dichro chloride:[12]

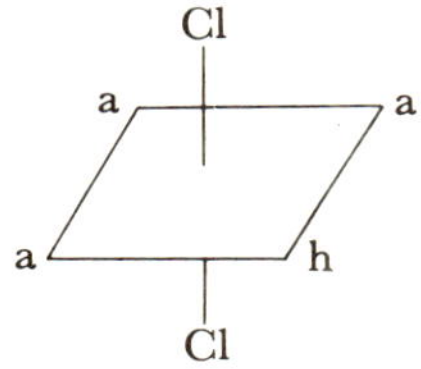

[6] [For example, $[Coen_2Cl_2]X$. Ammonia-violeo salts, *cis*-$[Co(NH_3)_4Cl_2]X$, whose existence was a necessary consequence of Werner's theory but not of Jørgensen's, were finally isolated by Werner in 1907 (*Ber.*, *40*, 4817–4825 (1907)). For a discussion and English translation see G. B. Kauffman, *Classics in Coordination Chemistry, Part I: The Selected Papers of Alfred Werner*, Dover Publications, Inc., New York, 1968, pp. 141–154). Their discovery convinced Jørgensen that his theory could not possibly be correct.—G.B.K.]

[7] *Z. anorg. Chem.*, *15*, 148 [143–172 (1897)].

[8] *Z. anorg. Chem.*, *8*, 183 [153–188 (1895)].

[9] [The compounds in question, $[Co(NH_3)_4(H_2O)Cl]X_2$, are now known to exist only in the *cis* form (violeo series).—G.B.K.]

[10] [That is, *trans*-(croceo) and *cis*-(flavo)$[Co(NH_3)_4(NO_2)_2]X$ (nitro) compounds, respectively.—G.B.K.]

[11] [Jørgensen regarded the croceo salts as nitro (—NO_2) compounds and the flavo salts as nitrito (—ONO) compounds, i.e., he considered them to be *structural* isomers. Werner, on the other hand, believed both salts to be *stereoisomers* containing nitro (—NO_2) groups; he correctly considered the flavo salts to be *cis* isomers and the croceo salts to be *trans* isomers.—G.B.K.]

[12] [*trans*-$[Co(NH_3)_3(H_2O)Cl_2]Cl$. For preparation see G. Schlessinger, *Inorg. Syn.*, *6*, 180–182 (1960).—G.B.K.]

corresponds completely to mine:

·Cl
Co·a·a·a·h·Cl
·Cl

he assumes[13] the formula:

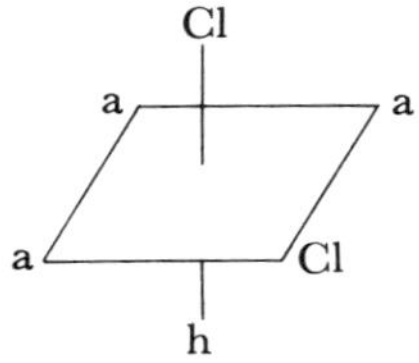

for an isomeric compound, whose identity, however, is still uncertain. Furthermore, it has not been investigated sufficiently for anything to be said about its constitution.[14] It is not even known whether it contains perhaps only one electronegative ion. It would be quite possible for it to be

·Cl
Co·a·a·a·Cl;
·OH_2·Cl

but even if a more exact investigation should show that 2 chlorine atoms are bonded to the cobalt in this compound, the isomerism could also be explained very well by the old theory, in which case the salt could be

·Cl
Co·a·a·h·a·Cl.
·Cl

It must be considered remarkable at least that, in spite of the energetic work which Werner and his students have devoted to the metal-ammonia salts, not one single compound has been found which, without any doubt, contains one chlorine atom or another electronegative radical in the corner of the square. If it is recalled how easily praseo salts are transformed into violeo salts and vice-

[13] *Z. anorg. Chem.*, *15*, 149 [143–172 (1897)].

[14] [The gray-green dichroic "dichro chloride" is the *trans* isomer, while the gray chloride is the *cis* isomer (K. Matsuno, *J. Coll. Sci. Imp. Univ. Tokyo*, *41*, Art. 10 (1921), p. 4; Y. Tanito, Y. Saito, and H. Kuroya, *Bull. Chem. Soc. Japan*, *25*, 328–331 (1952)).—G.B.K.]

versa in the ethylenediaminecobalt series,[15] it seems very questionable that in the ammonia series such a transition has never been successfully realized. Violeo salts isomeric with praseo salts are completely unknown in the ammonia series.[6]

With this conception, it is even less possible than was the case with the compounds mentioned under 1 and 2 above to fit the following compounds into the older system. These are compounds which, according to Werner, have no ions and thus have all their electronegative radicals in the corners of the square or octahedron. Here belong the dinitrotriamminecobalt salts,[16] the platosammine and platosemidiammine salts,[17] and the platinammine and platinsemidiammine salts.[18]

As for the dinitrotriamminecobalt salts, I must first emphasize that only the nitrite salt of this series, the so-called triamminecobalt nitrite, $(NO_2)_2{:}Co\cdot(NH_3)_3\cdot NO_2$, in agreement with Werner's theory, contains no ions, both according to its chemical and its physical-chemical behavior. Another salt of the series, the chloride, $(NO_2)_2{:}Co\cdot(NH_3)_3\cdot Cl$,[19] which, according to Werner's theory, should also contain no ions, is shown, both according to its chemical behavior and according to Werner's own rule of the conductivity in aqueous solution,[20] to contain an electronegative ion, and the experiments which he[21] has performed in order to support his theory are completely wrecked by Emil Petersen's[22] investigations. Whereas Werner's theory, however, as he himself admits, is not able to explain a compound $Co(NH_3)_3(NO_2)_2Cl$ with a negative ion, it is a different matter with the older theory with respect to the triammine nitrite. The latter is formed from the chloride by simple double decomposition with sodium nitrite. It is therefore obvious to assume that the actual dinitrotriammine nitrite which is thereby formed is

---

[15] [For example, the *trans* (praseo) compound may be converted into the *cis* (violeo) compound merely by heating, while treating the violeo compound with hydrochloric acid regenerates the original praseo compound.—G.B.K.]

[16] [$[Co(NH_3)_3(NO_2)_3]$. Jørgensen, however, regards this compound as a 1:1 electrolyte. The designation "salts" for the nonelectrolytes mentioned in this sentence is, of course, a misnomer.—G.B.K.]

[17] [*trans*- and *cis*-$[Pt(NH_3)_2Cl_2]$, respectively.—G.B.K.]

[18] [*trans*- and *cis*-$[Pt(NH_3)_2Cl_4]$, respectively.—G.B.K.]

[19] [$[Co(NH_3)_3(NO_2)_2Cl]$.—G.B.K.]

[20] [Jørgensen is mistaken. The compound, whose conductivity is determined in reference 21 is $[Co(NH_3)_4(NO_2)_2]Cl$, not $[Co(NH_3)_3(NO_2)_2Cl]$.—G.B.K.]

[21] [A. Werner and A. Miolati,] *Z. physik. Chem.*, *21*, 228 [225–238 (1896)].

[22] *Ibid.*, *22*, 414 [410–423 (1897)].

extraordinarily labile and is transformed immediately into the isomeric nitro compound, which cannot contain ions. Other possibilities, however, are not excluded. It is well known that in many nitrite double salts $2NO_2$ act as a divalent whole (unit), in which the nitrous acid has lost all or almost all its usual properties,[23] viz., as

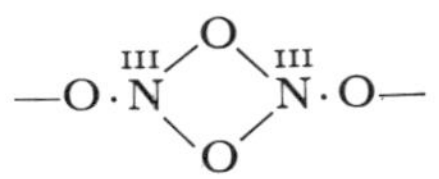 or as $-\mathrm{O}\cdot\overset{\mathrm{V}}{\mathrm{N}}\mathrm{O}=\overset{\mathrm{V}}{\mathrm{N}}\mathrm{O}\cdot\mathrm{O}-$

In agreement with this, the stable triammine nitrite could be explained easily as a compound without ions. Also all $3NO_2$ together could form a trivalent complex, 2 of whose valences would be bonded to cobalt and the third one to ammonia, as I have indicated earlier. Finally—and this possibility is perhaps the most probable—one can well imagine that an ammonium such as $(NO_2)_2{:}Co\cdot(NH_3)_3$ has so weak an electropositive character that the electrical opposition between it and such a weakly electronegative radical as NO_2 almost disappears. In this case, it is of course self-evident that the compound behaves as a nonconductor. This is unquestionably supported by the fact that the ammonium character of the radical $(NO_2)_2\cdot Co\cdot(NH_3)_2$ which is still poorer in ammonia has become so weakened that this radical even forms copulated [*gepaarte*] acids, e.g., $(NO_2)_2\cdot Co\cdot(NH_3)_2(NO_2)_2\cdot H$ and $(NO_2)_2\cdot Co\cdot(NH_3)_2\cdot C_2O_4\cdot H$, just like the weakly electropositive metals $\overset{\mathrm{III}}{\mathrm{Cr}}$, $\overset{\mathrm{III}}{\mathrm{Co}}$, $\overset{\mathrm{III}}{\mathrm{Ir}}$, and $\overset{\mathrm{III}}{\mathrm{Rh}}$.

Certainly the situation among the above-named platinum compounds is completely analogous. According to their entire chemical behavior, it is highly improbable that they do not contain ions, but, of course, it is sufficient that such ammonium radicals which contain only $2NH_3$ for every atom of platinum have very little distinct electropositive character and that their solutions exhibit little conductivity. As [Peter] Klason[24] has already emphasized, this cannot be

[23] See especially: [J.] Lang, "Om några nya Platinaoxydulföreningar," Upsala, 1861 (*J. prakt. Chem., 83,* 415 [(1861)]; L. F. Nilson, *N. Acta Universitat. Upsala,* [*15,* 55] (1877); A. Rosenheim and I. Koppel, *Z. anorg. Chem., 17,* 61 [35–72 (1898)].

[24] *Öfversigt af Kongliga Vetenskaps Akademiens Förhandlingar, Stockholm, 1895,* No. 5, 296 ff.; *Ber., 28,* 1487 [1477–1489 (1895)]. Klason's main argument against the theory of platinum bases, which I have defended, viz., that according to it only one platinumtriammonia chloride, $\mathrm{Pt}^{\cdot a\cdot a\cdot Cl,}_{\cdot a\cdot Cl}$ and only one platinummonoammonia chloride, $\mathrm{Pt}^{\cdot a\cdot Cl,}_{\cdot Cl}$ is possible, whereas according to his investigations "*in fact two platinumtriammonia chlorides exist and also in the monoammonias the ammonia*

surprising. Yet we see from Cossa's Salt[25] that even the atom group $Cl \cdot Pt \cdot NH_3 \cdot Cl$, which is even poorer in ammonia, quite analogously to mercuric chloride, is definitely inclined to form stable double salts with alkali metal chlorides, whereas the group $Cl \cdot Pt \cdot NH_3$ has almost completely lost its ammonium character. [N. S.] Kurnakow[26] has developed this manner of thinking in a very detailed and convincing manner in his interesting article on the complex metal bases,[27] and also he has specifically proved that the electropositive character of the ammonium radical is weakened not only by decreasing the number of ammonia groups but also by using less positive ammonias while retaining the number of these groups. As to the conductivity of these platinum-ammonia salts, incidentally, I wish to refer to the new observations and remarks of Emil Petersen cited below (p. 132).[28]

A series of remarkable ammonia compounds of trithiocarbonic acid [H_2CS_3] which were described by K. A. Hofmann,[29] partially in collaboration with Wiese,[30] seems to have made a not very small impact in favor of Werner's theory.[31] So it also seems to me, Hof-

---

can occupy two different positions," is no longer valid. Concerning the platinum-triammonia chloride, Klason had not noted that in the year 1894 (*Z. anorg. Chem.*, *7*, 316, Note) I already declared that Cleve's platomonodiammine chloride was possibly Cl·Pt·a·a·a·Cl; the true platomonodiammine chloride therefore remained to be discovered. Klason himself withdrew his assertion of a second platosemiammine chloride last year (*Z. anorg. Chem.*, *14*, 372, Note [1897]). Otherwise his experiments and views favor my theory more than Werner's theory. But I hope to refer to this on a later occasion.

[25] [Cossa's First Salt, potassium trichloroammineplatinate(II), $K[Pt(NH_3)Cl_3]$. —G.B.K.]

[26] [See G. B. Kauffman and A. Beck, *J. Chem. Educ.*, *39*, 44–49 (1962).—G.B.K.]

[27] *J. prakt. Chem.*, [2] *51*, 234 [(1895)], and *52*, 177[–192 (1895)].

[28] [Throughout this paper Jørgensen apparently refers to the page numbers of his original manuscript. Whenever possible, these have been changed to conform to the pages in this volume (*Classics, Part II*).—G.B.K.]

[29] *Z. anorg. Chem.*, *11*, 379[–384 (1896)], and *14*, 263[–281 (1897)].

[30] [The correct name is O. F. Wiede.—G.B.K.]

[31] See, e.g., [F.] Reitzenstein: *Ammoniak-, Pyridin-Salze und Hydrate bivalenter Metalle, Habilitationsschrift*, Universität Würzburg, 1898, p. 38. [Also see *Z. anorg. Chem.*, *18*, 253–304 (1896).—G.B.K.] Incidentally, I will take the liberty of correcting a few misunderstandings in this publication. If I said (*Z. anorg. Chem.*, *7*, 320 [289–330 (1894)] that Werner's conclusions that by the transformation of luteo into chloropurpureo chloride [$[Co(NH_3)_6]Cl_3$ into $[Co(NH_3)_5Cl]Cl_2$. —G.B.K.], etc., ammonia is *substituted* by chlorine, etc., is a subreption, then this is not, as Reitzenstein supposes (in his publication, p. 39), "*a very hard word.*" For subreption, as is well known, is in logic a certain kind of false inference and has here no offensive meaning whatsoever. [According to *Webster's Seventh New Collegiate Dictionary* (G. C. Merriam Co., Springfield, Mass.,

mann has showed very convincingly that these compounds do not contain ions, and this circumstance has induced him to agree with Werner's view. Metal-ammonia salts, without ions, however, are not completely inexplicable by the older theory. On the contrary, Hofmann's compounds can be explained quite easily by it, whereas they lead to very doubtful conclusions as far as Werner's theory is concerned. For what Hofmann has proved is just that his compounds do not contain ions, not that they do not contain ammonia "intercalated" ["*eingelagert*"] between the metal and the electronegative radical. These things are not identical. I will now try to show how such compounds can be explained by the older theory and the conclusions to which one is led by Werner's theory.

It seems quite common that a divalent acid radical, e.g., SO_4, C_2O_4, or SO_3, which is bonded simultaneously to ammonia and to a metal, loses completely by this bonding its character as an ion. Thirty years ago [P. T.] Cleve[32] prepared sulfatohydroxyloplatindiammine salts, whose rational formulas[33] were developed by [C. W.] Blomstrand[34] and in which the radical SO_4, although bonded to ammonia but also simultaneously to platinum, cannot be detected by barium salts. In the compound

$$\begin{array}{l} \quad\cdot O \\ \overset{}{_{IV}}\cdot NH_3\cdot NH_3\cdot O\cdot \end{array}\!\!\left.\begin{array}{l}\\ \\ \end{array}\right\} SO_2 \\ \begin{array}{l} Pt\cdot NH_3\cdot NH_3\cdot Cl \\ \quad\cdot OH \end{array}$$

1963), subreption is defined as "a deliberate misrepresentation" or "an inference drawn from it."—G.B.K.] For I did not at all mean that Werner was aware of this false inference since I explicitly added that he did not notice it. Also another remark which I made in the same article (p. 317), viz., that Werner's theory "has been received in many quarters with great approval—but of course hardly by those parties who have occupied themselves more extensively with metal-ammonia salts (because of a misprint, it says "has" ["*hat*"], but in the manuscript it says "have" ["*haben*"]), has been misinterpreted, not only by Reitzenstein (*loc. cit.*, p. 45) but also, and this I especially regret by Werner (*Z. anorg. Chem.*, *8*, 153[–188 (1895)]). What I intended to say was simply that none of those who had dealt experimentally more extensively with metal-ammonia salts (at that time there was mainly [F. A.] Genth, [O. W.] Gibbs, [P. T.] Cleve, F. Rose, [A.] Cossa, [G.] Vortmann, [C. W.] Blomstrand, and myself) had expressed himself in favor of Werner's theory.

[32] *Kongliga Svenska Vetenskaps Akademiens Handlingar, Stockholm, 7* (No. 6), 10 [1868].

[33] [Combinations of chemical symbols that indicate the way in which the atoms are probably linked in the molecule. They are less elaborate than structural formulas but more definite than empirical formulas.—G.B.K.]

[34] *[Die] Chemie der Jetztzeit [vom Standpunkte der electrochemischen Auffassung aus Berzelius Lehre entwickelt]*, Carl Winter's Universitäts-Buchhandlung, Heidelberg, 1869], p. 407.

the chlorine atom is definitely an ion and the SO_4 definitely a non-ion. Later I proved that similar conditions exist among the cobalt-ammonia salts. In the year 1885 I[35] found a series of sulfatopentamminecobalt salts of the general formula:

$$SO_4\begin{array}{l}\cdot NH_3\\ \;|\\ \cdot Co\cdot(NH_3)_4\cdot X,\end{array}$$

where *X* is an ion and SO_4 a non-ion.[36] The solution of the bromide is not precipitated by barium chloride.

A few years ago I[37] proved that there exists a completely analogous series of oxalopentamminecobalt salts:

$$C_2O_4\begin{array}{l}\cdot NH_3\\ \;|\\ \cdot Co\cdot(NH_3)_4\cdot X,\end{array}$$

in which the oxalic acid cannot be detected by calcium salts.[38] Completely analogous compounds, in which the sulfurous acid radical, SO_3, appears as a non-ion, viz.:

$$SO_3\begin{array}{l}\cdot NH_3\\ \;|\\ \cdot Co\cdot(NH_3)_4\cdot X,\end{array}$$

have been prepared recently by [A.] Werner and [H.] Grüger.[39,40] It is very interesting that these salts, exactly like my oxalopentamminecobalt salts, show a remarkable tendency to form acid salts with monobasic acids, e.g.,

$$HSO_3\cdot Co\begin{array}{l}\cdot NH_3\cdot Br\\ \cdot(NH_3)_4\cdot Br.\end{array}$$

Hofmann[40a] himself prepared the corresponding aquo compounds, e.g.,

$$SO_3\begin{array}{l}\cdot OH_2\\ \;|\\ \cdot Co\cdot(NH_3)_4\cdot X.\end{array}$$ [41]

[35] *J. prakt. Chem.*, [2] *31*, 262[–272 (1885)].

[36] $[[Co(NH_3)_5SO_4]X$. Here the sulfate group, although a dinegative anion, acts as a monodentate ligand.—G.B.K.]

[37] *Z. anorg. Chem.*, *11*, 418 [416–453 (1896)].

[38] $[[Co(NH_3)_5C_2O_4]X$. Also see footnote 36.—G.B.K.]

[39] *Ibid.*, *16*, 398[–423 (1898)].

[40] $[[Co(NH_3)_5SO_3]X$. Also see footnote 36.—G.B.K.]

[40a] [K. A. Hofmann and S. Reinsch,] *ibid.*, 377[–397 (1898)].

[41] $[[Co(NH_3)_4(H_2O)SO_3]X$.—G.B.K.]

From these he even obtained a nitrite salt in which neither NO_2 or SO_3 are ions and which therefore in all probability should be interpreted as:

$$\mathrm{NO_2\cdot Co\ \overline{\cdot(NH_3)_4\cdot}\ SO_3.}$$ [42]

Since the ammonia chain bonded to the divalent acid radical can thus consist of one, two, and four members, it must also be able to consist of three members, and Hofmann's compound $Co_2C_2S_7(NH_3)_6$,[43] in complete analogy with the above salts, will have to be formulated in a structural chemical manner as:

$$\mathrm{CS_3\ \overline{\cdot(NH_3)_3\cdot}\ Co\cdot S\cdot Co\ \overline{\cdot(NH_3)_3\cdot}\ CS_3.}$$

Such a compound should also not contain ions according to the older theory. In several respects, it recalls the indigo-blue oxalo chloride which I[44] obtained on treating dichrocobalt chloride with oxalic acid and which in all probability is

$$\begin{array}{l}\mathrm{Co\ \overline{\cdot(NH_3)_3\cdot}\ C_2O_4,\ \tfrac{1}{2}H_2O.}\\ \quad\mathrm{\cdot Cl}\end{array}$$ [45]

It also recalls the nitro oxalate:

$$\begin{array}{l}\mathrm{Co\ \overline{\cdot(NH_3)_3\cdot}\ C_2O_4}\\ \quad\mathrm{\cdot NO_2}\end{array}$$

obtained by Werner and [A.] Baselli[46] from a corresponding aquo salt,

$$\begin{array}{l}\mathrm{Co\ \overline{\cdot(NH_3)_3\cdot}\ C_2O_4\ .}\\ \quad\mathrm{\cdot OH_2\cdot NO_3}\end{array}$$ [47]

---

[42] $[[Co(NH_3)_4(SO_3)NO_2]$.—G.B.K.]

[43] [The original formula reads "$CoC_2S_7(NH_3)_6$."—G.B.K.]

[44] *Z. anorg. Chem.*, *11*, 334. [The correct reference is 434 (416–453 (1896)).—G.B.K.]

[45] $[[Co(NH_3)_3(C_2O_4)Cl]\cdot\frac{1}{2}H_2O$.—G.B.K.]

[46] *Z. anorg. Chem.*, *15*, 164 [143–172 (1897)]. [This compound is *cis*-$[Co(NH_3)_3(C_2O_4)NO_2]$. The *trans* form was also later prepared by Werner (*Ann.*, *405*, 212–241 (1914), p. 236).—G.B.K.]

[47] $[[Co(NH_3)_3(C_2O_4)H_2O]NO_3\cdot H_2O$. K. Matsuno (*J. Coll. Sci. Imp. Univ. Tokyo*, *41*, Art. 10 (1921), p. 25) later formulated it as a nonelectrolyte containing only one molecule of water, viz., $[Co(NH_3)_3(C_2O_4)NO_3]\cdot H_2O$.—G.B.K.]

It is self-evident that it will always be difficult to explain the constitution of such compounds because all the means which usually serve this end fail to work in exactly this situation. The analogy with the above-named compounds, which contain both ions and non-ions, strongly favors the view which has been cited. It deserves especially to be emphasized that Hofmann also succeeded in preparing a compound which corresponds to the above-mentioned acid salts, viz.:

$$\mathrm{HS\cdot Co}\begin{matrix}\cdot\overline{\qquad\quad}\cdot\\ \cdot(\mathrm{NH_3})_3\cdot\end{matrix}\mathrm{CS_3},$$

and which also does not contain ions since HS is bonded to the metal and CS_3 simultaneously to the metal and ammonia.[48] Also the composition of the rest of the compounds prepared by Hofmann, in which the valence of the metal is different, agrees completely with the view which I have developed here. Thus for divalent metals:

$$\mathrm{Ni}\begin{matrix}\cdot\overline{\qquad\quad}\cdot\\ \cdot(\mathrm{NH_3})_3\cdot\end{matrix}\mathrm{CS_3},\ \mathrm{Cu}\begin{matrix}\cdot\overline{\qquad}\cdot\\ \cdot\mathrm{NH_3}\cdot\end{matrix}\mathrm{CS_3},\ \mathrm{Zn}\begin{matrix}\cdot\overline{\qquad\quad}\cdot\\ \cdot(\mathrm{NH_3})_2\cdot\end{matrix}\mathrm{CS_3},\ \mathrm{Pt}\begin{matrix}\cdot\overline{\qquad\quad}\cdot\\ \cdot(\mathrm{NH_3})_2\cdot\end{matrix}\mathrm{CS_3}.$$ [49]

Thus no supposed arguments against the older theory can be deduced from Hofmann's interesting investigations. On the contrary, all his reliable and carefully investigated compounds can be explained in complete conformity with that theory. As has been adequately indicated above, the same is true of the cobalt-ammonia salts containing sulfurous acid, which were investigated by Werner and Grüger and simultaneously by Hofmann and Reinsch.

On the other hand, it is not clear to me how these compounds can be explained according to Werner's system. *Either* one must assume that the acid radicals CO_3, C_2O_4, SO_4, SO_3, and CS_3 occupy two coordination positions, e.g., in the carbonato-, oxalo-, and sulfitotetramminecobalt salts, and the fact that C_2O_4H and SO_3H in the acid pentammine salts occupy one coordination position favors this view; but according to this hypothesis the sulfato-, sulfito-, and oxalopentammine salts and also $Co(NH_3)_4(NO_2)SO_3$, $Ni(NH_3)_3(CS_3)$, and $Cu(NH_3)(CS_3)$ are unintelligible, for then $\overset{\text{III}}{\mathrm{Co}}$ certainly has the coordination number 7, $\overset{\text{II}}{\mathrm{Ni}}$ the coordination number 5, and $\overset{\text{II}}{\mathrm{Cu}}$ the coordination number 3, which certainly does not agree at all with

[48] [[$Co(NH_3)_3(CS_3)HS$].—G.B.K.]
[49] [[$M^{II}(NH_3)_nCS_3$].—G.B.K.]

the other ammonia compounds of these metals. *Or* one must assume that those acid radicals occupy only one coordination position, but then the carbonato-, oxalo-, and sulfitotetrammine salts and also $Co(NH_3)_3Cl(C_2O_4)$, $Co(NH_3)_3(NO_2)(C_2O_4)$, the dinitrodiamminecobalt oxalates, $Co(NO_2)_2(NH_3)_2C_2O_4M$ and further $Co_2(NH_3)_6S(CS_3)_2$, $Zn(NH_3)_2(CS_3)$, and $Pt(NH_3)_2(CS_3)$ become unintelligible since the coordination numbers then become 5 for Co, 3 for Zn, and 3 for Pt, which is likewise contrary to all experience. *Or* finally one must assume that the acid radicals in question occupy sometimes two and sometimes one coordination position, depending upon whether it agrees with Werner's theory or not.[50]

In Werner and [P.] Pfeiffer's[51] beautiful investigations of the behavior of tin alkyl halides toward ammonia and amines, we find just as little evidence which could weaken the older theory as in Hofmann's above-mentioned article. Compounds such as:

$$(C_2H_5)_2\cdot Sn\begin{matrix}\cdot NC_5H_5\cdot Cl\\ \cdot NC_5H_5\cdot Cl,\end{matrix}\qquad Cl_2\cdot \overset{IV}{Sn}\begin{matrix}\cdot S(CH_3)_2\cdot Cl\\ \cdot S(CH_3)_2\cdot Cl,\end{matrix}\qquad Cl_2\cdot Sn\begin{matrix}\cdot NC_5H_5\cdot Cl\\ \cdot NC_5H_5\cdot Cl,\end{matrix}$$

$$Br_2\cdot \overset{IV}{Sn}\begin{matrix}\cdot O(C_2H_5)_2\cdot Br\\ \cdot O(C_2H_5)_2\cdot Br,\end{matrix}\qquad (C_2H_5)_3\cdot Sn\cdot NC_6H_7\cdot J,^{52}\ \text{etc.,}$$

are explained by the older theory in complete analogy to the metal-ammonia salts. Indeed, in favor of the older theory while appearing questionable to the same extent for Werner's theory, he succeeded neither in finding the two isomers $(C_2H_5)_2\cdot Sn\cdot X_2$, which his theory requires,[53] nor in preparing addition products of tetraethyltin, tetraphenyltin, or tetraphenyllead, similar to the above. The latter fact seems to indicate that an electronegative radical is needed here to cause the intercalation [*Einlagerung*] of $\overset{V}{N}H_3$, $\overset{VI}{O}(C_2H_5)_2$, $\overset{IV}{O}H_2$, etc., which is assumed by the older theory. Meanwhile, it must be admitted that arguments for one theory or the other can scarcely be deduced from compounds, of which almost nothing is known but their empirical molecular formulas and their most

[50] [Actually, both Werner and Jørgensen are correct. A potentially bidentate ligand may sometimes be bidentate in one compound and monodentate in another.—G.B.K.]

[51] *Z. anorg. Chem.*, *17*, 82[–110 (1898)].

[52] [NC_6H_7 represents aniline, $C_6H_5NH_2$.—G.B.K.]

[53] [This is because tin(IV) is tetrahedral (*sp*^{3}) not square planar (*dsp*^{2}).—G.B.K.]

important physical properties, while chemical transformations, from which their constitution could be concluded, are completely lacking. For just this reason too, the disputes, by which Werner, from the existence of various compounds of boron fluoride, boron chloride, ethylboron, and methylboron, which are for the most part very little investigated and partially even doubtful, wants to conclude that "*the boron atom, which is saturated according to its valence number in the compounds* BR_3, *can bond still more atom groups, that is, more than this number permits us to predict*," these disputes mean so extraordinarily little for the question to be discussed here and all the less so, since all the compounds classified under the common name "addition products" may quite certainly not be interpreted from the same viewpoint. Moreover, [P. G.] Melikoff's and [L.] Pissarjewski's beautiful work on the perborate salts[54] has shown that the boron atom can probably occur with a valence higher than 3. In order not to ignore completely the compound of methylboron with ammonia,[55] which was emphasized so strongly by Werner, I wish to note that [E.] Frankland[56] indeed obtained ethylethylborate (*äthylborsaures Äthyl*) by limited action (otherwise ignition occurs) of oxygen on ethylboron.[57] $BE_3 + O_2 = B:E(OE)_2$. Thus the oxygen absorbed here is intercalated [*eingelagert*], and by analogy, the ammonia absorbed by ethylboron or methylboron with strong evolution of heat: $BM_3 + NH_3 = M_2B \cdot NH_3 \cdot M$, would also have to be considered as intercalated. Inasmuch as methylboron and ethylboron are very stable compounds, whereas BM_2H and BE_2H are in any case not known and perhaps not even capable of existing, it is not surprising that ethylboron and methylboron are re-formed from the ammonia compounds by action of acids or alkalis. On the other hand, the vapor density of methylboron-ammonia is abnormal. Just as $H \cdot NH_3 \cdot Cl$ is decomposed at high temperatures into $NH_3 + HCl$, so is methylboron-ammonia likewise decomposed into methylboron and ammonia. According to this, it would perhaps be more natural to assume that boron is pentavalent in the compound and that the compound itself is $H \cdot BM_3 \cdot NH_2$, which, in analogy with sal

---

[54] *Ber.*, *31*, 678[–680] and 983 [actually 953–956 (1898)].

[55] [$B(CH_3)_3 \cdot NH_3$—E. Frankland, *J. Chem. Soc.*, *15*, 363–381 (1862), p. 377; A. Stock and F. Zeidler, *Ber.*, *54*, 531–541 (1921), p. 539.—G.B.K.]

[56] *J. Chem. Soc.*, *15*, 383 (1862).

[57] E = ethyl, M = methyl.

ammoniac,[58] is decomposed into BM_3 and $H \cdot NH_2$. Incidentally, boron hydride also combines with ammonia.[59] Perhaps this would rather favor the latter view. [E.] Frankland also found that methylboron led into aqueous potassium carbonate forms potassium acid carbonate and the compound of methylboron with potassium hydroxide. According to this, methylboron has the character of an acid anhydride; indeed it even combines directly with potassium hydroxide to form $H \cdot BM_3 \cdot OK$, and just as $CO_2 + NH_3 = H \cdot CO_2 \cdot NH_2$, so would $BM_3 + NH_3 = H \cdot BM_3 \cdot NH_2$. Among such compounds, whose actual chemical character is difficult to decipher, much is conceivable, but extremely little is certain. One thing, however, seems certain, viz., that the existence of methylboron neither favors nor opposes the older theory of metal-ammonia salts.

Also the interesting investigations on complex [polynuclear] cobalt-ammonia salts which Werner and his students have published during the last years contain nothing that could weaken the older theory.

As for the oxy- and anhydrooxycobaltammine salts, I[60,61] had "obviously" ["*vor der Hand*"] viewed these compounds as:

$$\begin{array}{lr} HO\cdot & \cdot OH \\ X\cdot a_4\cdot \overset{IV}{Co}\cdot O\cdot \overset{IV}{Co}\cdot a_4\cdot X \\ X\cdot a\cdot & \cdot a\cdot X \end{array} \quad \text{and} \quad \begin{array}{lr} HO\cdot & \cdot X \\ X\cdot a_4\cdot \overset{IV}{Co}\cdot O\cdot \overset{IV}{Co}\cdot a_4\cdot X \\ X\cdot a\cdot & \cdot a\cdot X \end{array}$$

by changing the formulas of [G.] Vortmann,[62] to whom we are indebted for the more precise knowledge of these series of salts, only so far as to abandon the older and now scarcely tenable conception of cobalt atoms bonded directly to each other and to introduce a linkage of these atoms through oxygen. By their investigation of

---

[58] [NH_4Cl was often written as $\begin{matrix} H \\ H \end{matrix}\!>\!N\!\begin{matrix} \diagup H \\ -H \\ \diagdown Cl \end{matrix}$, and the nitrogen atom was regarded as quinquevalent at the time. See G. B. Kauffman, *Isis*, *63*, 78–95 (1973).—G.B.K.]

[59] [F.] Jones and [R. L.] Taylor, *J. Chem. Soc.*, *39*, 213[–219] (1881).

[60] [These compounds are now known to be (*a*) $[(NH_3)_5Co—O—O—Co(NH_3)_5]X_4$ and (*b*) $[(NH_3)_5Co—O—O—Co(NH_3)_5]X_5$, respectively. The former are brown, diamagnetic, and contain a peroxo (O_2^{2-}) bridge. The latter are green, paramagnetic, and contain a superoxo (O_2^{-}) bridge. Neither series contains cobalt(IV). See G. B. Kauffman, *Coord. Chem. Rev.*, *9*, 339–363 (1973), p. 347 ff.—G.B.K.]

[61] *Z. anorg. Chem.*, *14*, 411 [404–422 (1897)].

[62] *Sitzungsber. K. K. Akad. Wien* (*Math.-naturw. Kl.*), *91*, 1142 (1885).

these salts [A.] Werner and [A.] Mylius[63] have now reached mainly the same view as I, viz., (1) that the cobalt atoms are linked through oxygen; (2) that the compounds are "addition products of oxygen and pentammine salts"; (3) that in the anhydrooxy salts the atoms and atom groups bonded to cobalt are bonded to one cobalt atom as in the oxy salts and to the other as in the pentamminepurpureo salts. They proved the last point experimentally and, it appears to me, convincingly. It seems quite immaterial whether one now assumes for the oxy salts that both cobalt atoms are tetravalent, bonded by one oxygen atom and each holding a hydroxyl, as I have done, or whether one assumes with Blomstrand[64] that both cobalt atoms are tetravalent and linked through two single oxygen atoms, or whether one assumes with Werner that both cobalt atoms are trivalent and linked through two oxygen atoms which are bonded to each other, thus:

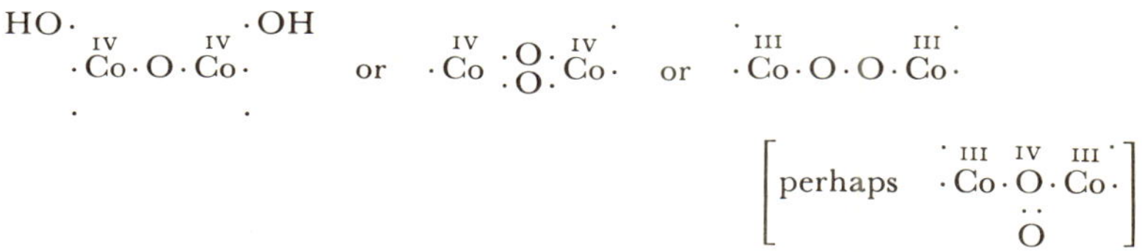

In all three cases, the quantity of available oxygen (i.e., oxygen more than in CoO) is that which was found by [E.] Fremy and [G.] Vortmann, viz., 1 atom O for 1 atom Co. The difference consists only in 1 molecule of water, which, however, according to Vortmann's analyses, is present in all these salts (Werner and Mylius did not analyze the oxy salts).[65] Almost the same holds true for the

---

[63] *Z. anorg. Chem.*, *16*, 248 [245–267 (1898)].

[64] [*Die*] *Chemie der Jetztzeit*, p. 291. [See reference 34—G.B.K.]

[65] Of the anhydrooxy salts, they analyzed the chloride and nitrate. In the first, they found 1 molecule of H_2O, just as Vortmann did. On the other hand, according to a cobalt and a nitrogen determination, they consider as anhydrous the nitrate, which according to Vortmann likewise contains 1 molecule of H_2O. I do not believe that this can be decided with certainty by use of such minimal amounts of substance (0.067 and 0.094 g.) as they have used. Yet Werner and Mylius are certainly correct. Many years ago I analyzed the salt, which had been dried over sulfuric acid (which thereby loses only very little weight) and obtained the same result:

0.4676 g. yielded 0.2300 g. $CoSO_4$ = 49.2% (calcd. for anhydrous salt 49.13; for the salt with 1 molecule of water 47.77).

0.5039 g., treated with barium hydroxide, etc., yielded 0.4679 g. $BaSO_4$ = 49.42% NO_3 (calcd. for anhydrous salt 49.13; for the salt with 1 H_2O 47.77).

anhydrooxy salts. Here too, the amount of available oxygen, though theoretically not the same yet very close to that found by Vortmann, is according to Werner's and my formulas, viz.,

$$\begin{array}{l} \mathrm{HO}\cdot\ \ \overset{\mathrm{IV}}{}\qquad\ \ \ \ \overset{\mathrm{IV}}{}\cdot\mathrm{X} \\ \mathrm{X}\cdot\mathrm{a_4}\cdot\mathrm{Co}\cdot\mathrm{O}\cdot\mathrm{Co}\cdot\mathrm{a_4}\cdot\mathrm{X} \\ \mathrm{X}\cdot\mathrm{a.}\qquad\qquad\ \ \cdot\mathrm{a}\cdot\mathrm{X} \end{array} \quad \text{and} \quad \mathrm{X_2}\cdot\mathrm{a_5}\cdot\overset{\mathrm{III}}{\mathrm{Co}}\cdot\mathrm{O}\cdot\mathrm{O}\cdot\overset{\mathrm{IV}}{\mathrm{Co}}\begin{array}{l}\cdot\mathrm{X} \\ \cdot\mathrm{a_5}\cdot\mathrm{X_2},\end{array}$$

(Blomstrand, when he wrote his book,[34] did not know of this series of salts), and the analytical difference (1 atom of hydrogen to 30) falls entirely within the experimental error. However, since according to Vortmann the anhydrooxy salts are formed from oxy salts by action of acids with formation of water (which is also indicated by their name), I believe that my formulas are the most probable. For with Werner's formulas this transition is unintelligible and also one of the cobalt atoms becomes 4-valent. But it *must* become tetravalent, otherwise the atom group Co·O·O·Co cannot, according to Werner's own rule,[66] become 5-valent, and there is much in favor of one X's being bonded to cobalt.

For the older theory of metal-ammonia salts, both Werner's and my formulas are formally acceptable, but for Werner's theory, neither of the two is really acceptable. For according to my formulas both cobalt atoms in both series of salts have the coordination number 7, while according to Werner's theory, this is true of one cobalt atom in the anhydrooxy salts.

The investigations which Werner, in collaboration with [F.] Beddow, [A.] Baselli, and [F.] Steinitzer,[67] has carried out on the constituents of Vortmann's fusco sulfate,[68] are of very great interest. Here they were concerned with complex cobalt-ammonia salts of an entirely new type, and the perseverance and skill with which they prepared these very difficultly accessible compounds and explained their relationship, are truly admirable. But it is a pity that the analyses were carried out with such small amounts of substance.

[66] *Z. anorg. Chem.*, *3*, 323 [267–330 (1893)]. [This, Werner's first paper on the coordination theory, is discussed and translated into English in Kauffman, *Classics*, pp. 5–88.—G.B.K.]

[67] *Z. anorg. Chem.*, *16*, 109[–166 (1898)].

[68] [A mixture of 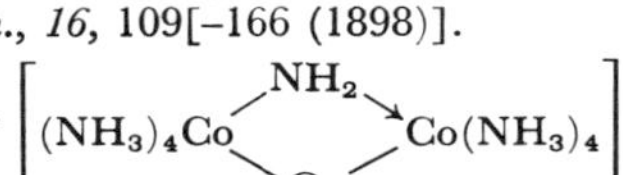 $(SO_4)_2$,

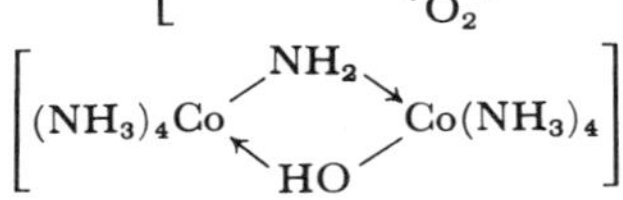 $(SO_4)_2$, and other materials.—G.B.K.]

Never 0.2, usually less than 0.1, often less than 0.05, indeed even 0.04 g. were used, whereby indeed at best the validity of whole percentage numbers is guaranteed, and every systematic error in the method, each experimental error, each impurity of the substance or of the reagents exerts an undue influence. Fortunately, there is usually no doubt about the relative number of the different atoms, but sometimes there is an uncertainty about the presence or absence of 1 molecule of water. Thus 1 molecule of H_2O is assumed in the acid hydrosulfato-imidooctammine sulfate, but the determinations of Co, N, and H correspond better to $2H_2O$, and the determination of SO_4 just as well to the anhydrous salt. For the chloride nitrate of the same series it is found that the salt which was dried over sulfuric acid loses 2 and 2.7% of its weight at 80° and 100°, respectively. In spite of this, the composition is referred to the salt dried over sulfuric acid, and the salt is considered to be anhydrous. For the nitrate of the same series, which was analyzed only once in the pure form, the hydrogen which was found corresponds better to a salt with $1H_2O$ than to an anhydrous salt. Conversely, the percentage found for SO_4 is 1.2% higher than that for an anhydrous salt. Now, although the figures found for Co and N correspond quite well to an anhydrous salt, yet, because of the small amounts of substance used for the analyses, the question can be raised whether the salt nevertheless does contain 1 molecule of H_2O, which would amount to only 2.8% distributed over five constituents. In any case, it seems very risky to conclude from this with Werner that the salt is anhydrous and that the two cobalt atoms cannot be linked through oxygen but must be bonded through imide in the series of salts in question as well as in the imidooctammine salts.

If we did not have better reasons, the hydrosulfatonitrate could just as well be:

$$(NO_3)_2 \cdot a_6 : Co \cdot O \cdot Co \begin{matrix} \cdot SO_4H \\ \cdot a_3 \cdot NO_3 \end{matrix} \quad \text{as} \quad (NO_3)_2 \cdot a_5 : Co \cdot NH \cdot Co \begin{matrix} \cdot SO_4H \\ \cdot a_3 \cdot NO_3, \end{matrix}$$

and the imidooctamminecobalt chloride could just as well be:

$$Cl_2 \cdot a_6 : Co \cdot O \cdot Co \begin{matrix} \cdot Cl \\ \cdot a_3 \cdot Cl \end{matrix}, 4H_2O \quad \text{as} \quad Cl_2 \cdot a_5 : Co \cdot NH \cdot Co \begin{matrix} \cdot Cl \\ \cdot a_3 \cdot Cl \end{matrix}, 5H_2O$$

where the distribution of the ammonia molecules is surely arbitrary (as in the formulas given below).

But the composition of the hydronitrito-imidooctammine salts decidedly favors the imide bonding. Despite the fact that they all have 1 molecule of water, they cannot contain $O + 9NH_3$. For this is surely indicated by the behavior of this series of salts, viz., that the NO_2 group, which is contained in it, is bonded to cobalt. Likewise, everything favors both cobalt atoms' being trivalent in this series of salts. But neither according to the older theory nor according to Werner's rule can an atom group $:Co \cdot O \cdot Co:^{NO_2}$, with or without ammonia, bond more than three equivalents of an electronegative radical. But this series of salts contains four. Hence it definitely follows that the cobalt atoms must be bonded through the imide group. For in the imide group, the valence of the nitrogen can be increased to five[69] in the presence of strong acids, which must always be present in the preparation of these salts:

$$\cdot\overset{\text{III}}{N}H\cdot + HX = \cdot\underset{\underset{HX}{\wedge}}{\overset{\text{V}}{N}}H\cdot,$$

and thereby the constitution of the hydronitro-imidooctammine salts can be explained quite simply in analogy with the rhodo- and erythrochromium salts.[70]

According to this, the hydronitro-imidooctamminecobalt salts[71] become:

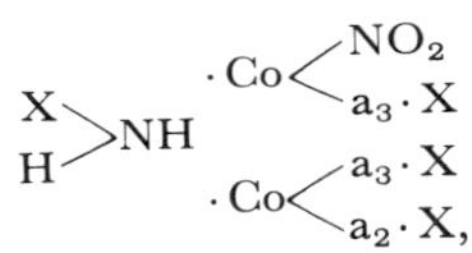

[69] [In terms of modern orbital theory, quinquevalent or pentavelent nitrogen is, of course, impossible, since elements of the second period are limited to a covalence or coordination number of four. The Blomstrand–Jørgensen theory, however, is based on chains containing quinquevalent nitrogen, and nitrogen(V) appears frequently in this paper.—G.B.K.]

[70] [S. M. Jørgensen,] *J. prakt. Chem.*, [2] *45*, 279 [260–280 (1892)].

[71] [Werner[67] first formulated these salts as

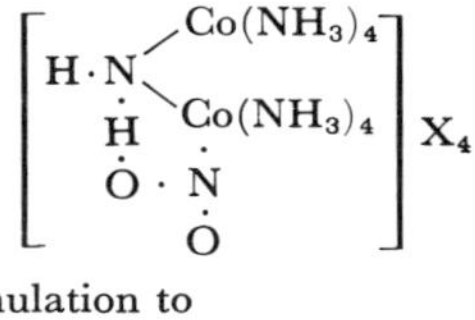

but later changed his formulation to

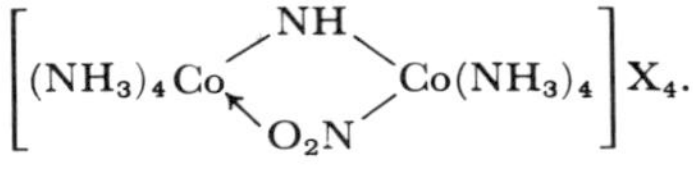

[72 (pp. 14, 54, 56, 58–60)]

—G.B.K.]

and the hydrosulfato-imidooctammine salts[73] quite analogously become:

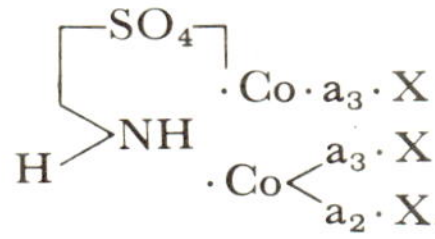

If I have understood Werner correctly, this view concerning the bonding of the sulfuric acid radical coincides with the view which he assumes. It is essentially the same as the one which is assumed by the older theory in order to explain the sulfato-, oxalo-, and sulfitopentammine salts.[75] The method of explanation, however, presupposes a divalent acid radical. $(SO_4)^{II}$ substitutes here for $NO_2 + X$ in the hydronitro-imidooctammine salts.[71] Werner's formula for the latter:[76]

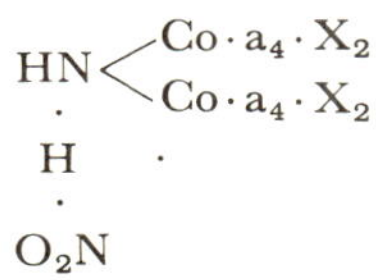

is absolutely unintelligible to me. I suspect that the formula should be interpreted as:

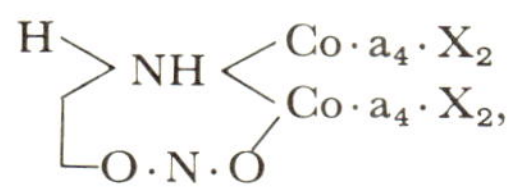

---

[72] [A. Werner, *Ann.*, *375*, 1–144 (1910).—G.B.K.]

[73] [Werner[67] first formulated these salts as

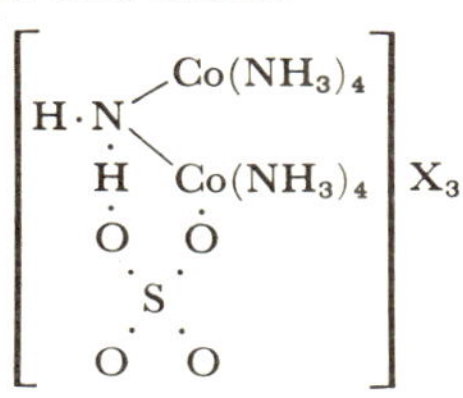

but later changed his formulation to

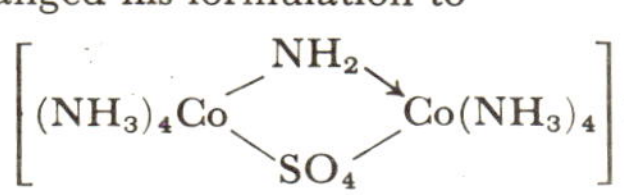

—G.B.K.]

[74] [A. Werner, *Ber.*, *40*, 4605–4615 (1907).—G.B.K.]

[75] See above p. 7 ff. [p. 111 ff.[28]—G.B.K.]

[76] [A. Werner, F. Beddow, A. Baselli, and F. Steinitzer,] *Z. anorg. Chem.*, *16*, 116 [109–166 (1898)].

but then the salts are derived from an unknown dibasic nitrogen acid, HO·N·OH with divalent nitrogen, and they have nothing in common with the numerous nitro and nitrito compounds which are usually well known among the cobalt-ammonia salts. It would be desirable if Werner had expressed himself more extensively here. For I also cannot see the analogy with the diamminecobalt nitrites (even according to his view).

I think that it must be admitted that the older theory explains the relationships both more simply and more naturally than does Werner's theory, all the more so since the hydrosulfato-[73] and hydronitro-imidooctammine[71] salts thereby become completely analogous to the other imidooctammine salts, a fact which I now intend to show.

On warming the chloride nitrate of the hydrosulfato-imidooctammine[73] series with hydrochloric acid and precipitating with ethyl alcohol, a product was obtained, which after being washed free of acid with ethyl alcohol, being dissolved in water, and being precipitated with hydrochloric acid, furnished little, shiny, blood-red prisms.[77] Surely, these too were washed free of acid with ethyl alcohol, although this is not mentioned explicitly. This salt was supposedly imidooctammine chloride, $NH{:}Co \cdot a_8Cl_4 \cdot 5H_2O$, and the determination of Co, N, and H also correspond quite well, but the hydrogen determination (by elementary analysis) came out too low. On the other hand, the quantity of chlorine was found to be 6.3% too high. "Since the too high chlorine content leads one to suspect a mechanical contamination of the chloride," it was recrystallized from water, and now a similar salt was obtained, whose analysis agrees with the above formula of the chloride. Now the only compound which could influence only the chlorine determination but not the others is dry hydrochloric acid,[78] and a contamination of the salt with 6 to 7% of it is absolutely inconceivable.[79] All

[77] [Reference 76, p. 150 ff.—G.B.K.]

[78] [Hydrogen chloride.—G.B.K.]

[79] [Werner originally (1898)[77] formulated this compound as

$$\left[\begin{matrix}(NH_3)_4 \\ X\end{matrix} Co—NH—Co \begin{matrix}(NH_3)_4 \\ X\end{matrix}\right] X_2,$$

but later[72] reformulated it as

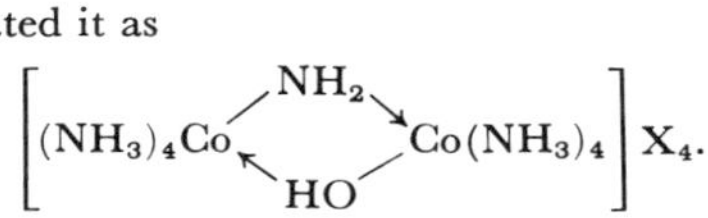

this therefore seems to me to indicate to a high degree that the original crystals were an acid salt of the formula:

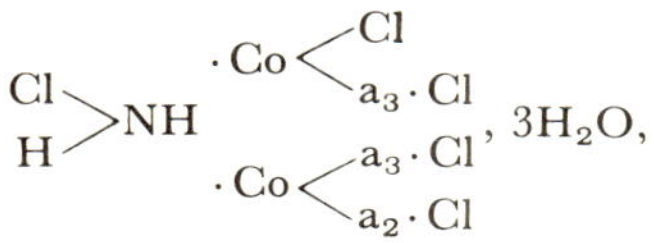

consequently completely analogous to hydrosulfato- and hydronitroimidooctammine salts. For this, then, the following is calculated:

	[Sum of Atomic Weights]	[Percentages] [Calcd:]	Found:
2Co	118	23.5	23.2
9N	126	25.2	25.4
5Cl	177.5	35.4	34.3
32H	32	6.4	6.8
3O	48	9.5	—
	501.5	100.0	

By recrystallization from pure water this salt would then lose 1 molecule of HCl and be transformed into the ordinary imidooctammine chloride.[79] That one of the chlorine atoms in the latter salt is bonded to cobalt is therefore probable inasmuch as the nitrate can be obtained free of chlorine only on treating the chloride with hot silver nitrate solution. Whether other salts too with 5 equivalents of an electronegative radical can be prepared must remain uncertain. Those which were also investigated were analyzed only

In 1907 he reported an alternative preparation for the tetrahydrated chloride, which gave much better analytical results for chlorine (28.34% found; 28.30% calculated).[74] In his long paper of 1910 Werner[72(p. 44)] again commented that the chlorine content was too high. Probable reasons for the discrepancies have been recently suggested by B. Barro, R. E. Marsh, and W. P. Schaefer (*Inorg. Chem.*, *9*, 2131–2137 (1970)). They believed Werner's compounds of 1898 and 1910 were

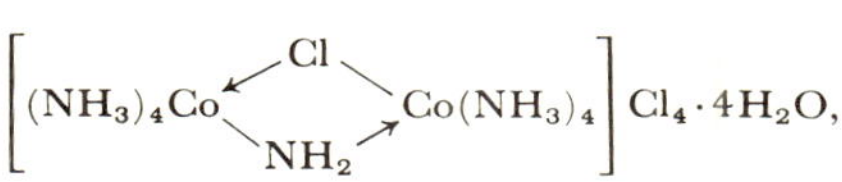

the first dicobalt species shown to contain a chloro bridge, and his compound of 1907 was the true chloride of the μ-amido-μ-hydroxo series. The two compounds are closely related, and the latter can be prepared from the former by standing in aqueous solution.—G.B.K.]

after recrystallization from or washing with water. It therefore seems possible that they too can produce acid salts from sufficiently acidic solutions. In any case, the fact that this series of salts, as Werner[80] reports, can be converted by concentrated sulfuric acid into the hydrosulfato series is in favor of this view.

The three series of salts mentioned thus far are all prepared directly or indirectly from the so-called oxo-diimidooctammine salts.[81] By reduction with sulfurous acid the oxo-diimido nitrate is transformed into the hydrosulfatoimido nitrate, which again yields the imidooctammine chloride on warming with hydrochloric acid. The oxo-diimido salts on simultaneous reduction and nitration with nitrous acid are likewise transformed into the hydro-nitro-imidooctammine salts, which can also be obtained from the imidooctammine salts with nitrous acid. From this Werner concludes: (1) that the oxo-diimido salts contain their ammonia bonded in the same way as the hydrosulfato-, hydronitro-, and imidooctammine salts; (2) that in the oxo-diimido salts at least one of the cobalt atoms is tetravalent; (3) that, since the oxo-diimido salts contain only 4 equivalents of an electronegative radical for every 2 atoms of Co, one of the valences of the tetravalent cobalt atom must be bonded to oxygen:

$$:\overset{\text{III}}{\mathrm{Co}}\cdot\mathrm{NH}\cdot\overset{\text{IV}}{\mathrm{Co}}\cdot \\ \cdot\mathrm{O}.$$

Temporarily no objection shall be made to this. Taking all this into consideration, Werner regards the following formulas as being

---

[80] [A. Werner, F. Beddow, A. Baselli, and F. Steinitzer,] *Z. anorg. Chem.*, *16*, 115 [109–166 (1898)].

[81] [Werner first formulated these compounds as

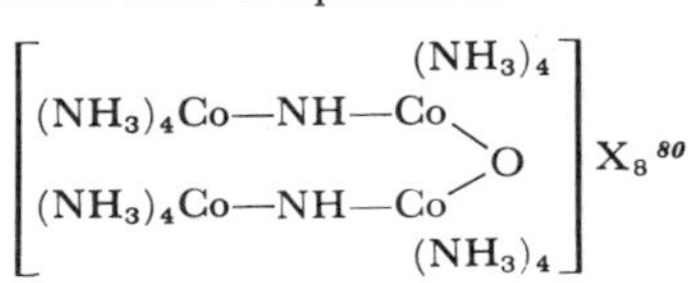

but later reformulated them as

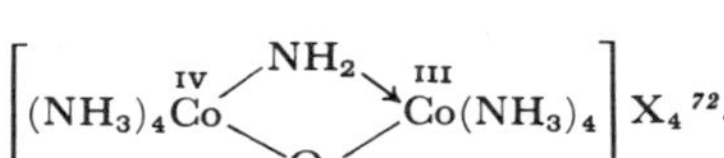

The compounds contain a superoxo (O_2^-) bridge rather than cobalt(IV).—G.B.K.]

the most probable for the oxo-diimido- and imidooctammine chloride:

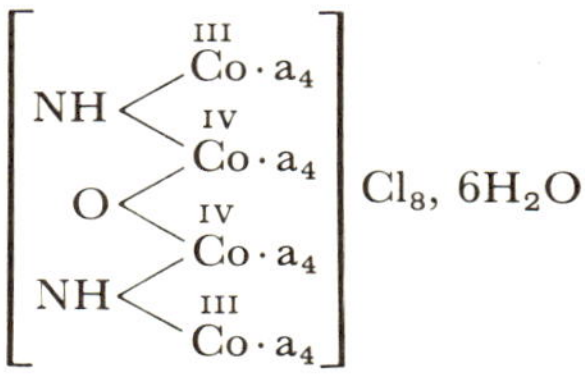

or

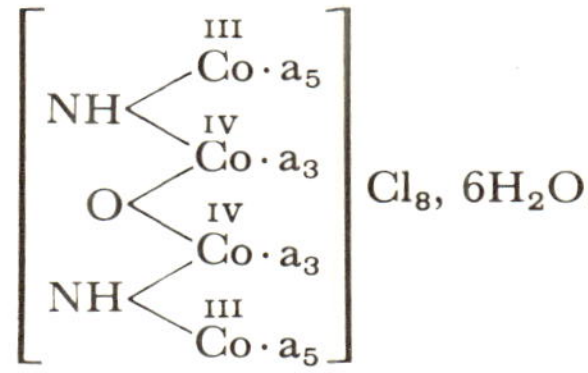

and

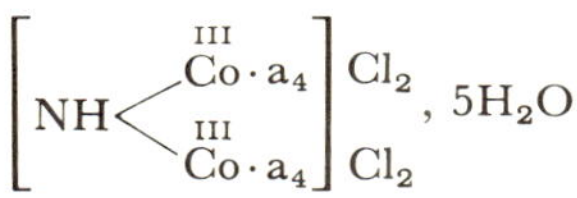

or according to the facts cited above perhaps even better:

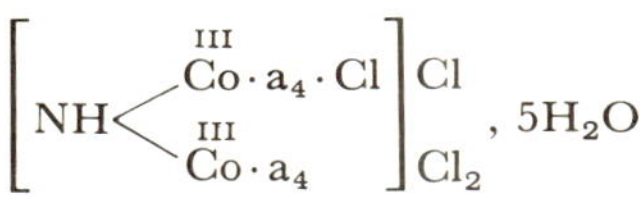

It is at once striking that in the first formulas 2 and in the latter one of the cobalt atoms appears with coordination number 5. However, no importance can be given to this, since all these salts contain water and it is quite possible that 2 or 1 molecules of water, respectively, are bonded differently from the others. But if the above formulas are compared with those of rhodochromium chloride[82,83] and rhodosocobalt chloride,[84,85] which according to this are analogous, viz. (written according to Werner's theory):

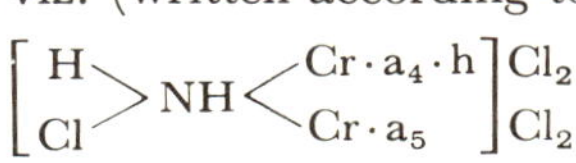

and

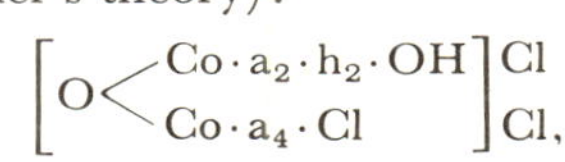

[82] [S. M. Jørgensen,] *J. prakt. Chem.* [2], *25*, 330 [321–346 (1882)].

[83] [[$(NH_3)_5CrOHCr(NH_3)_5$]X_5.—G.B.K.]

[84] [S. M. Jørgensen,] *Z. anorg. Chem.*, *16*, 188 [184–197 (1898)].

[85] [This compound, originally called anhydrobasic tetrammine-diaquodiammine-cobalt chloride by Jørgensen, was later shown by Werner, E. Berl, E. Zinggeler, and G. Jantsch (*Ber.*, *40*, 2103–2125 (1907), p. 2118) to be the tetranuclear complex

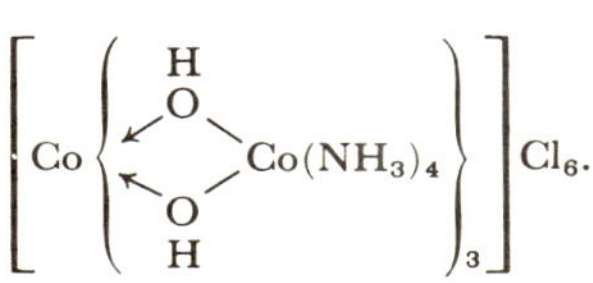

then it must seem quite strange that none of his new compounds yields chloropentammine chloride or chloroaquotetrammine chloride on being heated with hydrochloric acid. For rhodochromium chloride is thereby converted smoothly into 2 molecules of chloropentammine chloride, and half of the rhodosocobalt chloride, which is constructed analogously to imidooctammine chloride, is converted just as smoothly into chloroaquotetrammine chloride. Indeed even the compound

$$\left[O<\begin{matrix}Co\cdot a_4\cdot Cl\\ Co\cdot a_4\cdot Cl\end{matrix}\right]\begin{matrix}Cl\\ Cl,\end{matrix}$$

which is formed on treating oxo-diimidooctammine chloride[81] with ammonium carbonate and which is constructed exactly like one of the halves of the rhodosocobalt chloride, can be heated with hydrochloric acid without the separation of chloroaquotetrammine chloride. This is absolutely inconceivable according to Werner's theory, for according to his theory all the ammonias which are coordinated with one metal atom are bonded in exactly the same manner. For Werner's theory, the difference which the older theory assumes between the four-membered ammonia groups and the ammonia molecules which stand alone does not exist. But the behavior of Werner's new salts which has been mentioned above is easily explained by the older theory. For just this reason I have never introduced the four-membered ammonia group into those formulas which are not written according to Werner's system. After all, we are concerned here with compounds of a completely new type, which do not contain four-membered ammonia groups at all.

Besides, according to Werner's formulas, however, only one half of the oxo-diimidooctammine nitrate[81] can be transformed by sulfurous acid into hydrosulfato-imido nitrate; the other half must remain behind as imidooctammine nitrate:

(See G. B. Kauffman, *Coord. Chem. Rev.*, *9*, 339–363 (1973), p. 357.) It is an irony of history that the bromide of this series, first prepared by Jørgensen, became the first completely carbon-free coordination compound to be resolved, thus unequivocally proving Werner's octahedral configuration for cobalt(III) (A. Werner, *Ber.*, *47*, 3087–3094 (1914). For a discussion and English translation see Kauffman, *Classics*, pp. 174–184).—G.B.K.]

$$\begin{array}{ll} & \overset{\mathrm{III}}{\mathrm{Co}}\cdot \mathrm{a_4}\cdot(\mathrm{NO_3})_2 \\ \mathrm{NH}< & \\ & \overset{\mathrm{IV}}{\mathrm{Co}}\cdot \mathrm{a_4}\cdot(\mathrm{NO_3})_2 \\ \mathrm{O}< & \\ & \overset{\mathrm{IV}}{\mathrm{Co}}\cdot \mathrm{a_4}\cdot(\mathrm{NO_3})_2 \\ \mathrm{NH}< & \\ & \overset{\mathrm{III}}{\mathrm{Co}}\cdot \mathrm{a_4}\cdot(\mathrm{NO_3})_2 \end{array} + \mathrm{H_2SO_3} = \left[\mathrm{NH}<\begin{array}{l}\mathrm{Co}\cdot \mathrm{a_4}\cdot\mathrm{SO_4H}\\ \mathrm{Co}\cdot \mathrm{a_4}\end{array}\right]\begin{array}{l}\mathrm{NO_3}\\ (\mathrm{NO_3})_2\end{array} + \mathrm{HNO_3} + \left[\mathrm{NH}<\begin{array}{l}\mathrm{Co}\cdot \mathrm{a_4}\\ \mathrm{Co}\cdot \mathrm{a_4}\end{array}\right]\begin{array}{l}(\mathrm{NO_3})_2\\ (\mathrm{NO_3})_2.\end{array}$$

Thereupon nitric acid oxidizes the sulfurous acid:

$$\left[\mathrm{NH}<\begin{array}{l}\mathrm{Co}\cdot \mathrm{a_4}\\ \mathrm{Co}\cdot \mathrm{a_4}\end{array}\right]\begin{array}{l}(\mathrm{NO_3})_2\\ (\mathrm{NO_3})_2\end{array} + \mathrm{H_2SO_3} = \left[\mathrm{NH}<\begin{array}{l}\mathrm{Co}\cdot \mathrm{a_4}\cdot\mathrm{SO_4H}\\ \mathrm{Co}\cdot \mathrm{a_4}\end{array}\right]\begin{array}{l}\mathrm{NO_3}\\ (\mathrm{NO_3})_2\end{array} + \mathrm{HNO_2}.$$

In general, it seems peculiar that Werner[86] speaks of the "surprisingly easy conversion of oxo-diimido salts into hydrosulfato salts by the action of sulfurous acid," for one can see from the experimental part of his article[87] that the product has "not yet been investigated." It "resembles strikingly the nitrate of the sulfato series and also contains sulfuric acid"; nothing further is known of it. Only one thing is certain—that on suitable treatment with hydrochloric acid it is converted into imidooctammine chloride.[79] Therefore, only the fact that the oxo-diimido-[81] and the imidooctammine[79] salts contain their ammonia molecules bonded in the same manner has been proven experimentally. But it is not known whether the sulfurous acid acts other than as a reducing agent on the nitric acid. But thereby the only reason for assuming tetravalent cobalt in this series of salts becomes quite uncertain. The behavior of the salts themselves does not indicate this. While the anhydrooxycobaltammine salts[60b]—not to speak of the oxy salts[60a]—decompose almost immediately in pure aqueous solution with evolution of oxygen, the aqueous solution of oxo-diimido nitrate decomposes quickly to be sure "with an odor of ammonia and precipitation of cobaltic oxide," but nothing is said about evolution of oxygen. It is even reported of a nitrate with a higher water content that "when a *warm* solution of it is mixed with nitric acid," the nitrate containing less water is precipitated. This does not indicate tetravalent cobalt. To be sure, the presence of tetravalent cobalt in the oxo-diimido salts[81] is possible, but not certain or even probable.

---

[86] [A. Werner, F. Beddow, A. Baselli, and F. Steinitzer,] *Z. anorg. Chem.*, *16*, 117 [109–166 (1898)].

[87] [A. Werner, F. Beddow, A. Baselli, and F. Steinitzer,] *Z. anorg. Chem.*, *16*, 138 [109–166 (1898)].

I now want to show how easily everything can be explained according to the older theory. By the assumptions, which, according to the data cited above, are indeed probable, viz., that the oxo-diimido salts[81] contain no four-membered ammonia groups at all and no tetravalent cobalt at all, a much simpler formula can be assigned to this series of salts, viz., one in which even the imide group is bonded to acid, i.e.,

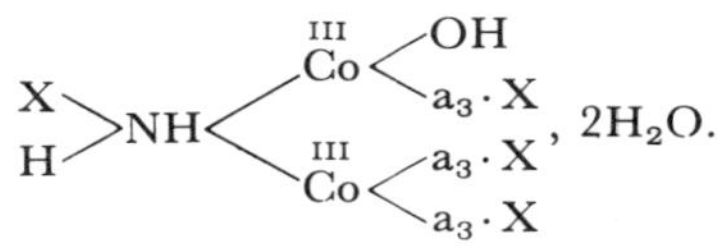

This formula agrees with the figures found by analysis just as well as Werner's formula, sometimes a little better, sometimes a little worse. The distribution of the two three-membered and of the one two-membered ammonia groups is, of course, arbitrary, but the ammonia is bonded exactly as in the formula given for the imidooctammine salts.[79] With this constitution of the oxo-diimido salts,[81] the two components, which Werner so beautifully proved to be contained in Vortmann's fusco sulfate, i.e., the red imidooctammine sulfate[68,79] and the green oxo-diimido sulfate,[68,81] become isomeric,[88] viz.:

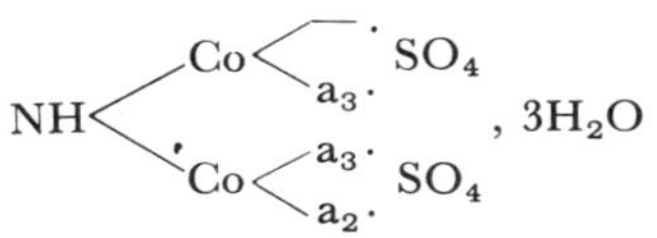

and

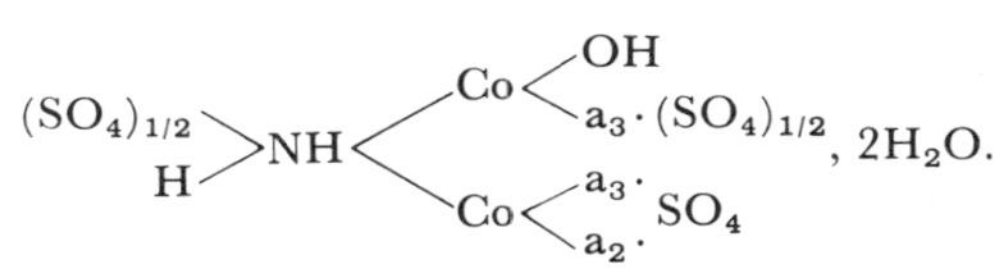

With nitric acid, the first compound yields hydrosulfatooctammine nitrate:

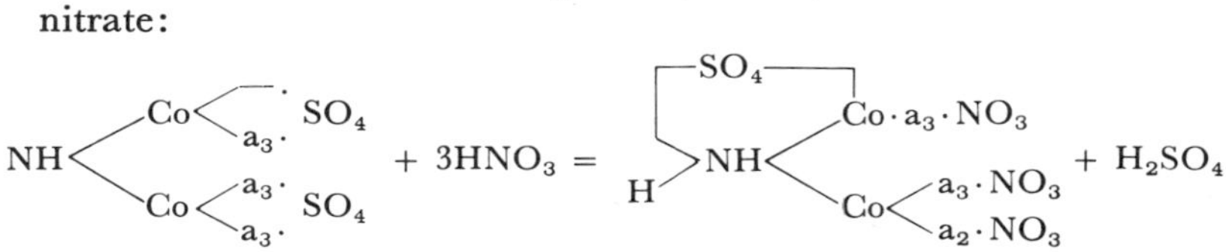

[88] See the analytical results summarized in [A. Werner, F. Beddow, A. Baselli, and F. Steinitzer,] *Z. anorg. Chem.*, *16*, 112 [109–166 (1898)].

By simple double decomposition with nitric acid, the other compound yields oxo-diimido nitrate.[81] With nitrous acid, this simply yields hydronitrito-imido nitrate:

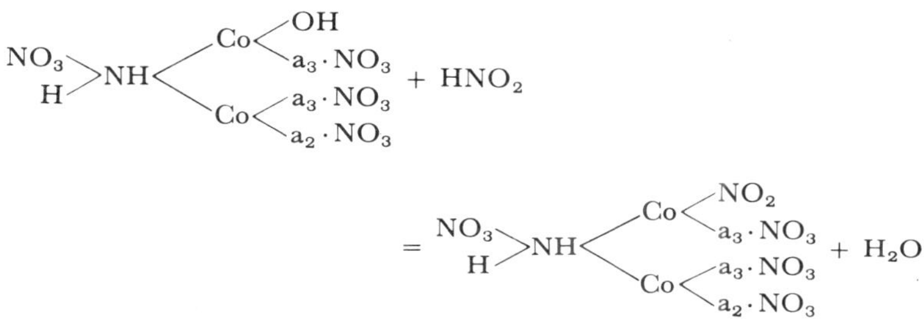

With sulfurous acid, the oxo-diimido nitrate[81] yields either hydrosulflto or hydrosulfato-imido nitrate,[73] without or with reduction of the nitric acid, respectively:

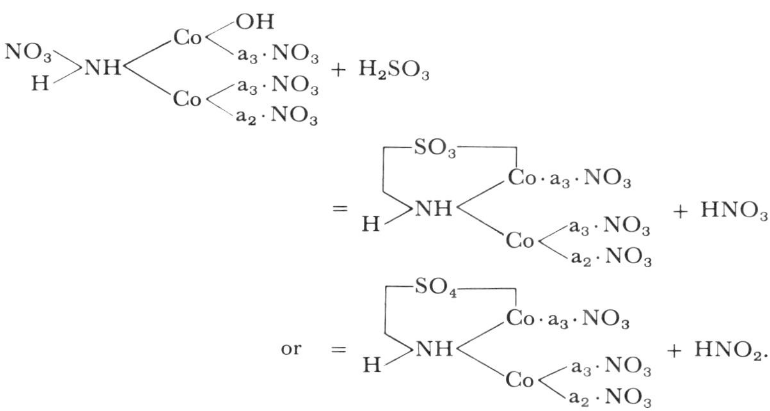

According to the data cited above (p. 124),[28] with hydrochloric acid, both compounds probably first yield acid

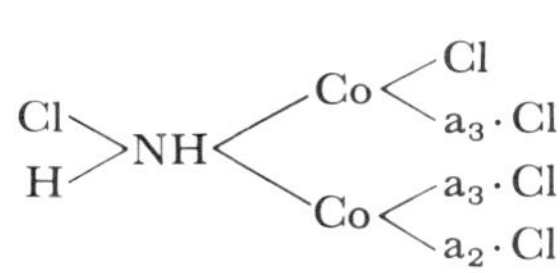

and then with splitting off of hydrogen chloride they yield normal [neutral] imidooctammine chloride:[79]

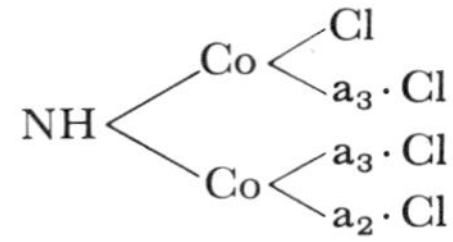

If treatment of oxo-diimido nitrate[81] with ammonium carbonate, followed by treatment with hydrochloric acid, yields anhydrobasic octammine chloride, then one has:

(1) 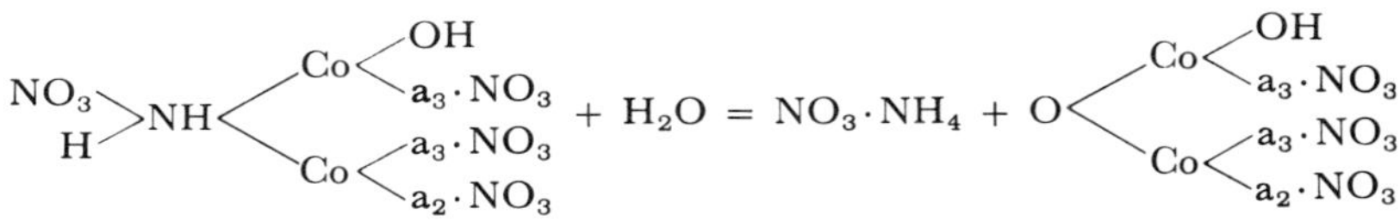

(2) 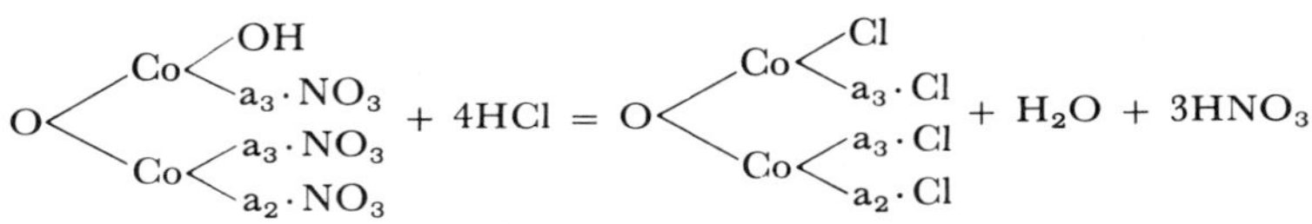

Thus everything is explained in the simplest manner, especially the fact that none of the compounds yields chloropentammine or chloroaquotetrammine chloride on heating with hydrochloric acid.

If the imide of the true imidooctamminecobalt salts[79] binds acid only exceptionally and probably only in strongly acidic solutions, while the rhodo-[83] and erythrochromium[89] salts not only always contain acid bonded to imide but even by the action of alkalies they form basic salts, in which the acid radical bonded to imide is replaced by hydroxyl, then I can hardly attribute, with Werner, this different behavior to a "hidden cause" ["*causa occulta*"] in the metals cobalt and chromium. It appears more likely to me that the entire radical of the rhodo[83] and erythro salts[89] receives, from the presence of the twice-present four-membered ammonia group, a more electropositive character than here, where we are dealing only with two- and three-membered ammonia groups. Therefore those imidooctammine salts[79] which contain only the weakly electronegative radicals OH and NO_2 are more inclined to bind acid to imide than those which contain the strongly electronegative radicals Cl and NO_3 bonded to metal. Only if a divalent acid radical can unite simultaneously with imide and metal, as in the hydrosulfatoimido salts,[73] do such imidocobalt salts also have acid bonded to

---

[89] [$[(NH_3)_5CrOHCr(NH_3)_4H_2O]X_5$, first prepared by Jørgensen (*J. prakt. Chem.* [2], *25*, 398–430 (1882)).—G.B.K.]

imide. It therefore might be possible that the conjugated nitric acid $H \cdot (NO_2)_2 \cdot H$[90] could form such salts. Perhaps the compound which according to Baselli[91] is formed by use of sodium nitrite and acetic acid is just such a salt. The great family resemblance of the melano salts,[92] which according to the older theory also contain the three-membered ammonia group, to the imidooctammine salts[79] deserves to be emphasized. It too is in favor of the presence of the same group in the latter [imidooctammines].

With isolated exceptions, which are discussed below in another context, I have considered above almost all the latest works on metal-ammonia salts, and I have examined these results in the light of the older theory and of the Werner theory as objectively as was possible for me, but I have thereby found no reason for abandoning the older theory. On the contrary, in some essential points it seems to me to possess no small advantages. In the following, I shall treat several situations of a general nature, where it seems to me that Werner's theory exhibits considerable weakness as compared to the older theory.

Werner early attributed great importance to the conductivity

[90] See p. 108 above.[28]

[91] [A. Werner, F. Beddow, A. Baselli, and F. Steinitzer,] *Z. anorg. Chem.*, *16*, 157 [109–166 (1898)].

[92] [The chloride is a black, difficultly-soluble precipitate obtained by addition of concentrated hydrochloric acid to an ammoniacal cobalt(II) solution partially oxidized by air (F. Rose, *Untersuchungen über ammoniakalische Kobaltverbindungen*, Carl Winter, Heidelberg, 1871, p. 39). G. Vortmann proposed the name *Melanochlorid* (*Ber.*, *10*, 1451–1459 (1877), p. 1455; *ibid.*, *15*, 1890–1903 (1882), p. 1901). It consists of

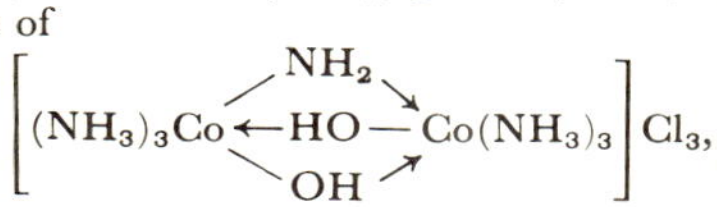

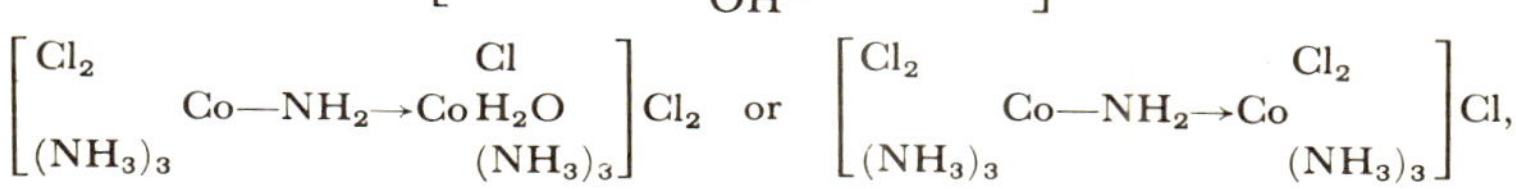

and 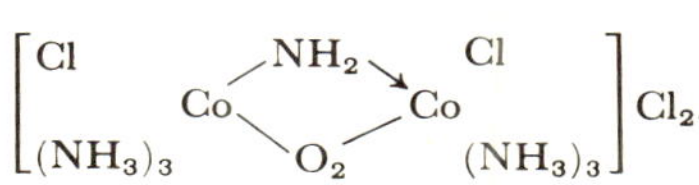 .—G.B.K.]

of metal-ammonia salts. Indeed, his first experimental work on these compounds[93] was the determination of their conductivity. In his theoretical explanations[94] he always refers to the fact that these investigations provided us with information about the number of ions present in the metal-ammonia salts and thereby furnished "a direct determination of the constitution" of these compounds. Earlier[95] I protested in a joking manner against this representation of the situation, and here I seriously repeat my protest. For the great and preponderant number of cobalt-, chromium-, and rhodium-ammonia salts, I had earlier shown by chemical methods not only the number of ions but also *which ions these were*, in part many years before Werner ever occupied himself with this question.[96] According to the nature of things, the electrical conductivity can give no information at all concerning the last-mentioned question; this question can only be answered by chemical means, and only this answer can give information about the constitution of the compound. On the other hand, with respect to *the number of ions*, Werner and Miolati's conductivity determinations have led to an empirical rule,[97] which in most cases agrees remarkably well with the results found previously by chemical means. How purely empirical this rule nevertheless is can best be seen from the fact that it does not at all agree with the number of ions actually present in aqueous solutions of these compounds. As is well known, Emil Petersen[98] found the latter number by an unobjectionable method, viz., by determining both the osmotic pressure of the solution by means of the freezing point depression as well as by its conductivity, and both quantities as far as possible under the same external conditions. It was then shown that the number of ions actually present in the solutions is entirely different from the number to which Werner and Miolati's

---

[93] [A. Werner and A. Miolati,] *Z. physik. Chem.*, *12*, 35[–55] (1893). [For a discussion and English translation see Kauffman, *Classics*, pp. 89–115. —G.B.K.]

[94] See, e.g., [A. Werner and P. Pfeiffer,] *Z. anorg. Chem.*, *17*, 82 ff. [–110 (1898)].

[95] *Ibid.*, *7*, 317 [289–330 (1894)].

[96] [Jørgensen's first publication on coordination compounds (chloropurpureo-cobalt salts, $[Co(NH_3)_5Cl]X_2$) appeared in 1878 (*J. prakt. Chem.*, [2] *18*, 209–247 (1878)), when Werner was only eleven years old!—G.B.K.]

[97] *Z. physik. Chem.*, *14*, 507 [506–521 (1894)]. [For a discussion and English translation see Kauffman, *Classics*, pp. 117–139.—G.B.K.]

[98] *Z. physik. Chem.*, *22*, 410[–423 (1897)].

empirical rule had led. Thus the number of ions is as follows for:

According to Werner and Miolati:[99]		$\mu_{25°}$, $v = 1000$	According to Petersen:
$(NO_2)_2 \cdot Co \cdot a_3 \cdot Cl$	2	115.8	3
$(NO_2)_2 \cdot Co \cdot a_4 \cdot NO_3$ (Flavo)[10]	2	106.2	4
$(NO_2)_2 \cdot Co \cdot a_4 \cdot NO_3$ (Croceo)[10]	2	98.3	4
$Cl_2 \cdot Co \cdot a_4 \cdot Cl$[6]	3–4	364.7	4
$Cl_2 \cdot Co \cdot en_2 \cdot Br$ (Praseo)[100]	2	135.3	3
$Br_2Co \cdot en_2 \cdot NO_3$ (Praseo)	3?	221?	4
$Cl_2 \cdot Co \cdot en_2 \cdot Cl$ (Violeo)	2–3	175.7	4
$Cl_2 \cdot Co \cdot en_2 \cdot NO_3$ (Violeo)	2–3	177.1	4
$(NO_2)Cl \cdot Co \cdot a_4 \cdot Cl$	3	251.1	4
$CO_3 \cdot Co \cdot a_4 \cdot Br$	2	115.2	3

Afterward, at my request, Emil Petersen investigated the conductivities of several other metal-ammonia salts, all of which I myself prepared and analyzed. I am taking the liberty of reporting his results here, all the more since he has added to them some remarks of considerable interest.

1. *Dichro chloride*, $Cl_2 \cdot Co \cdot a_3 \cdot h \cdot Cl \cdot$[12]—Pure recrystallized[101] salt was used for the experiments. Because of the instability of the aqueous solutions,[102] a new solution was prepared for each of the investigated concentrations, by means of which the freezing point and the conductivity were determined as quickly as possible. All the figures found therefore represent initial values:

---

[99] [The modern formulas for the following compounds are: $[Co(NH_3)_3(NO_2)_2Cl]$, *cis*- and *trans*-$[Co(NH_3)_4(NO_2)_2]NO_3$, *trans*-$[Co(NH_3)_4Cl_2]Cl$, *trans*-$[Coen_2Cl_2]Br$, *trans*-$[Coen_2Br_2]NO_3$, *cis*-$[Coen_2Cl_2]Cl$, *cis*-$[Coen_2Cl_2]NO_3$, *trans*-$[Co(NH_3)_4(NO_2)Cl]Cl$, and $[Co(NH_3)_4CO_3]Br$, respectively. Werner and Miolati accounted for those cases in which the conductivities were higher than theoretically required by postulating aquation reactions, e.g.,

$$[Co(NH_3)_3(NO_2)_2Cl] + H_2O \rightarrow [Co(NH_3)_3(NO_2)_2(H_2O)]^+ + Cl^-.$$

—G.B.K.]

[100] en = ethylenediamine.

[101] [S. M. Jørgensen,] *Z. anorg. Chem.*, *17*, 475–476 [455–479 (1898)].

[102] [According to Werner and Miolati, dichrocobalt chloride is converted into triammineroseocobalt chloride by aquation: $[Co(NH_3)_3(H_2O)Cl_2]Cl + 2H_2O \rightarrow [Co(NH_3)_3(H_2O)_3]Cl_3$.—G.B.K.]

v	μ_{0°[103]
25	127.4
50	140.2
100	156.9
200	180.8
1600	219.5
3200	221.8

v	Observed depression	Molecular depression	i	$1 + (k - 1)\alpha$ $k = 5$
25	0.260	65.0	3.48	3.30
50	0.140	70.0	3.74	3.53
100	0.075	75.0	4.01	3.83
200	0.038	76.0	4.06	4.26

Thus it appears that the salt in aqueous solution contains not less than 5 ions. However, it must be remembered that the determination of the maximal conductivity ($v = 3200$) and therewith the calculation of the degree of dissociation (α) is somewhat uncertain (i.e., μ_∞ rather too high) because of the instability of the salt, especially in dilute solution. In any case, the experiments show that the dissolved salt is decomposed into at least 4 ions.

2. *Platosemidiammine chloride*, $Pt\begin{smallmatrix}\cdot a \cdot a \cdot Cl \\ \cdot Cl\end{smallmatrix}$[104]

v		μ_{25°[103]
1000	Initial value	5.4
	After 15 min.	8.6
	25 ,,	9.2
	45 ,,	12.8
	90 ,,	20.8

3. *Platosemidiethylenediamine chloride*, $Pt\begin{smallmatrix}\cdot en \cdot Cl \\ \cdot Cl\end{smallmatrix}$[105]

[103] The values found agree approximately with those of Werner and Miolati (*Z. physik. Chem.*, *14*, 518 [506–521 (1894)). See Kauffman, *Classics*, p. 135.—G.B.K.]

[104] [Peyrone's salt, *cis*-[$Pt(NH_3)_2Cl_2$]. For preparation and properties see G. B. Kauffman and D. O. Cowan, *Inorg. Syn.*, *7*, 239–245 (1963).—G.B.K.]

[105] [[$PtenCl_2$].—G.B.K.]

v		$\mu_{25°}$
1000	Initial value	44.0
	After 12 min.	44.6
	90 ,,	47.0
	150 ,,	55.2

4. *Platinsemidiammine chloride*, $Cl_2 \cdot Pt \begin{matrix} \cdot a \cdot a \cdot Cl \\ \cdot Cl \end{matrix}$ [106]

v	$\mu_{25°}$	
125	16.9	After 3 hours 36.1
250	20.6	
500	24.0	
1000	27.3	

With another solution there was found:

		Petersen:		Werner and Miolati:
v		$\mu_{25°}$		$\mu_{25°}$
1000	Initial value	25.6		6.99
	After 5 min.	32.0	After 2 min.	9.02
	15 ,,	39.6	4 ,,	11.04
	30 ,,	53.3	7 ,,	14.62
	60 ,,	79.2	10 ,,	16.63
			15 ,,	20.52

E. Petersen notes that at the concentration $v = 125$ some salt crystallizes out at the freezing point of the solution. Therefore the freezing point could not be determined. For the concentration $v = 250$, the depression $= 0.014°$ was found, a number which cannot make any great claim to accuracy. According to this, the molecular depression would be $= 24.0$, and from this i is calculated to be $= 1.28$, thus considerably higher than for a nonconductor, which indeed agrees quite well with the conductivity which was found. The latter was found by E. Petersen to be somewhat higher than that found by Werner and Miolati, as is evident from the above data. But even greater is the difference between Emil

[106] [*cis*-[$Pt(NH_3)_2Cl_4$]. For preparation and properties see G. B. Kauffman, *Inorg. Syn.*, *7*, 236–238 (1963).—G.B.K.]

Petersen's and Werner and Miolati's conductivity determinations for:

5. *Platinammine chloride*, $Cl_2Pt\cdot\begin{matrix}\cdot a\cdot Cl\\ \cdot a\cdot Cl\end{matrix}$ [107]

Here Werner and Miolati found:

v		μ_{25°
1000	Initial value	3.52
	After 10 min.	14.92
	20 ,,	23.88
	60 ,,	33.40

On the other hand, Petersen found:

v	μ_{25°	
125	135.2	After 2½ hours 135.3
250	142.3	
500	155.5	
1000	180.0	
2000	204.9	
4000	242.3	

With a directly prepared solution he found:

v		μ_{25°
1000	Initial value	62.0
	After 10 min.	92.7
	20 ,,	111.8
	35 ,,	145.7
	60 ,,	187.7
	240 ,,	318.2

With a third solution there was found:

v	μ_0		
250	116.5	After 4 hours 103.6	After 6½ hours 101.3
500	123.5		
1000	128.5		
2000	128.6		
4000	128.6		

The freezing point depression was determined with the same

---

[107] [*trans*-[$Pt(NH_3)_2Cl_4$]. For preparation and properties see G. B. Kauffman, *Inorg. Syn.*, *7*, 236–238 (1963).—G.B.K.]

solution as the conductivity at 0° and, like the latter, as soon as possible after preparing the solution. The following was found:

v	Observed depression	Molecular depression	i	$1 + (k - 1)\alpha$ for $k = 4$
250	0.028	70.0	3.74	3.72

According to this the number of radicals acting as ions in the molecule should be 4. In addition, concerning the meaning of the experimental results for the question of constitution, E. Petersen now notes the following:

"The equation

$$i = 1 + (k - 1)\alpha$$

or the calculation of the number of ions from

$$k = \frac{i - 1}{\alpha} + 1$$

presupposes that all the atoms or radicals functioning as ions are separated *to the same degree*. Indeed, however, this need not necessarily be the case if $k > 2$, and it is probably not so in the present case. Inasmuch as the dissociation proceeds in the following manner:

$$Cl_2 \cdot Pt \cdot a_2 \cdot Cl_2 = [Cl_2 \cdot \overset{++}{Pt} \cdot a_2] + \bar{Cl} + \bar{Cl}$$

and then:

$$Cl_2 \cdot Pt \cdot a_2 = [Cl \cdot \overset{+}{Pt} \cdot a_2] + \bar{Cl},$$

and inasmuch as both processes are complete (i.e., the degree of dissociation in both processes = 1), the absolute number of ions per unit volume will become 4 times the number of molecules and the freezing point depression will become 4 times the normal value ($i = 4$)."

"However, if the dissociation proceeds in the following manner:

$$Cl_2 \cdot Pt \cdot a_2 \cdot Cl_2 = [Cl_2 \overset{+}{\cdot} Pt \overset{+}{\cdot} a_2] + \bar{Cl} + \bar{Cl}$$

and afterward:

$$Cl_2 \cdot Pt \cdot a_2 = [\overset{+}{Pt} \cdot \overset{+}{a_2}] + \bar{Cl} + \bar{Cl},$$

and if the first process is complete, and the other one proceeds only half way (i.e., if the degree of dissociation of the latter $= \frac{1}{2}$), then the absolute number of ions per unit volume will obviously become the same as above, viz.,

$$2n + \frac{n}{2} + \frac{3n}{2} = 4n,$$

therefore the freezing point depression will also be the same, while, nevertheless, each molecule in the first case forms 3 chlorine[108] ions + the complex $[\mathrm{Cl}\cdot\mathrm{Pt}\cdot\mathrm{a}_2]$ and in the latter, on the other hand, 3 chlorine[108] ions + the two positive complexes, $[\mathrm{Cl}_2\cdot\mathrm{Pt}\cdot\mathrm{a}_2]$ and $[\mathrm{Pt}\cdot\mathrm{a}_2]$." [109]

"A similar analysis can be carried out for the cobalt compounds investigated earlier. If *M* designates the positive complex and X a negative atom or radical, then a compound $\mathrm{X^I}_2\cdot\mathrm{M}\cdot\mathrm{X}$ can of course be imagined to dissociate in the following manner:

$$\mathrm{X^I}_2\cdot\mathrm{M}\cdot\mathrm{X} = [\mathrm{X^I}_2\overset{+}{\cdot}\mathrm{M}] + \overset{-}{\mathrm{X}}$$

and

$$\mathrm{X^I}_2\cdot\mathrm{M} = [\overset{+}{\mathrm{X^I}}\cdot\mathrm{M}] + \overset{-}{\mathrm{X^I}}."$$

"The magnitude of *i*, however, could thus never become greater than 3, a value which would be attained by complete dissociation according to both processes. On the other hand, if the dissociation were to proceed in the following manner:

$$\mathrm{X^I}_2\cdot\mathrm{M}\cdot\mathrm{X} = [\mathrm{X^I}_2\overset{+}{\cdot}\mathrm{M}] + \overset{-}{\mathrm{X}}$$

and then:

$$\mathrm{X^I}_2\cdot\mathrm{M} = \overset{++}{\mathrm{M}} + \overset{-}{\mathrm{X^I}} + \overset{-}{\mathrm{X^I}},$$

then (assuming that the first process were complete) the degree of dissociation of the second process would be

$$\begin{array}{ll} >\frac{1}{2} \text{ yields: } & i > 3, \\ =\frac{1}{2} & i = 3, \\ <\frac{1}{2} & i < 3." \end{array}$$

[108] [Chloride.—G.B.K.]

[109] Since the migration velocity can be different for the radicals, which are different in the two cases, the conductivity need not necessarily be the same, but the difference will probably not be very great.

"For three of the cobalt-ammonia salts investigated, i was found to be <3[110] and for the rest >3." (See p. 134 above).

"It therefore appears to me that it can indeed be assumed that the degree of dissociation is, as a rule, the same for both radicals X^I but different from that of the radical X, corresponding to the special and usually equal manner of bonding of the first two radicals. For compounds where $i > 3$, this can probably be combined with the assumption that all 3 negative radicals are separated to the same degree; indeed, the greater probability favors the supposition expressed above that the first radical is split off completely even in concentrated solution, whereas the other two are partially split off later by dilution and during standing."

"On the contrary, the experimental results seem to be completely inconsistent with Werner's theory."

To me too, this way of explaining the experimental results seems as interesting as it is probable. In any case, the state of affairs does not at all seem to favor Werner's theory. Among the metal-ammonia salts, there are some, which according to Werner's theory *must* necessarily be nonconductors and can contain no ions at all. The compounds in question are: triamminecobalt nitrite,[111,112]

---

[110] For carbonatotetrammine bromide i can be only <3.

[111] [Gibbs' salt or Erdmann's triammine, $[Co(NH_3)_3(NO_2)_3]$, discovered by O. L. Erdmann (*J. prakt. Chem.*, *97*, 412 (1866)) and later investigated by W. Gibbs (*Proc. Am. Acad.*, *10*, 14 (1875)). A particularly crucial compound in the controversy between Jørgensen and Werner, the large number of "polymerization isomers" of empirical formula $Co(NH_3)_3(NO_2)_3$ resulted in uncertainties as to which compounds were actually the monomolecular isomers *cis*- and *trans*-$[Co(NH_3)_3(NO_2)_3]$, predicted by Werner's coordination theory. Inasmuch as the existence of two and only two nonelectrolytic isomers of zero conductivity were required by Werner's theory, the confusion in this case enabled Jørgensen to attack the new theory. In fact, the uncertainties were not cleared up until very recently. According to T. E. MacDermott and S. Barfoed (*Acta Chem. Scand.*, *24*, 924–930 (1970)), the various claims for the isolation of the *facial* (1,2,3) isomer of $[Co(NH_3)_3(NO_2)_3]$ required by Werner's theory are all without foundation, and the compound most commonly assigned this configuration is actually *trans*-$[Co(NH_3)_4(NO_2)_2]$*trans*-$[Co(NH_3)_2(NO_2)_4]$ (M. LePostollec and J.-P. Mathieu, *Compt. rend.*, *257*, 3185 (1963)). Only the *meridional* (1,2,6) isomer has been identified (S. M. Jørgensen, *Z. anorg. Chem.*, *17*, 455–479 (1898), p. 475; Y. Tanito, Y. Saito, and H. Kuroya, *Bull. Chem. Soc. Japan*, *25*, 188–191 (1952)).—G.B.K.]

[112] Werner earlier (*Z. anorg. Chem.*, *8*, 174 ff. [153–188 (1895)]) assumed two modifications for this compound, and from this isomerism he drew far-reaching conclusions. After I thoroughly investigated the question (*ibid.*, *13*, 175 ff. [172–190 (1897)]), he seems to have abandoned this isomerism. At least he no longer mentions two triammine nitrites in compilations of nonconductors ([A. Werner and P. Pfeiffer,] *ibid.*, *17*, 84 [82–110 (1898)]).

dinitrotriamminecobalt chloride,[113,114] platosemidiammine chloride,[104] platosemidiethylenediamine chloride,[105] platosammine chloride,[115] platinsemidiammine chloride,[106] and platinammine chloride.[107] Now of course the experiments have shown that triamminecobalt nitrite[111] and platosemidiammine chloride[104] are nonconductors. On the other hand, according to E. Petersen's experiments cited above, platosemidiethylenediamine chloride,[105] platosammine chloride,[115] and platinsemidiammine chloride[106] exhibited a not quite minimal conductivity. According to Petersen's results, platinammine chloride[107] and dinitrotriamminecobalt chloride[113] are definitely good conductors. Of the 7 compounds, therefore, 2 are nonconductors, 3 are poor conductors, and 2 are good conductors. This state of affairs seems to be very questionable for Werner's theory, especially in view of the great importance which he [Werner] has attributed to the conductivity values.

As has already been emphasized above (p. 126),[28] in Werner's system all the ammonia molecules which are bonded to one metal atom are bonded in exactly the same manner. Also, all the electronegative radicals, which he imagines to be situated at the corners of the octahedron, are bonded in exactly the same manner. But I think that the actual circumstances definitely indicate that there are appreciable differences both between the ammonia molecules and also between the coordinated electronegative radicals.

Thus, according to Werner, all $6NH_3$ are bonded in exactly the same manner in the hexammine salts Ma_6X_3.[116] There is therefore also no difference at all between the three electronegative ions. According to my theory, on the other hand, these salts are

---

[113] [$[Co(NH_3)_3(NO_2)_2Cl]$. For aquation reaction see footnote 99.—G.B.K.]

[114] Werner earlier declared in very positive terms ([A. Werner and A. Miolati,] *Z. physik. Chem.*, *21*, 224 [1896]) [this should read p. 225, not p. 224.—G.B.K.] that this salt was a nonconductor, as it *must* be according to his theory. Now, after Emil Petersen's refutation (*ibid.*, *22*, 413 ff. [410–423 (1897)]), he seems to have abandoned this assumption. At least he no longer cites it among the nonconductors ([A. Werner and P. Pfeiffer,] *Z. anorg. Chem.*, *17*, 84 [82–110 (1898)]).

[115] [Reiset's Second Chloride, *trans*-$[Pt(NH_3)_2Cl_2]$. For preparation and properties see G. B. Kauffman and D. O. Cowan, *Inorg. Syn.*, *7*, 239–245 (1963).—G.B.K.]

[116] [$[M(NH_3)_6]X_3$, the so-called luteo salts.—G.B.K.]

$$\begin{array}{l} \quad \cdot a \cdot X \\ M \cdot a \cdot a \cdot a \cdot a \cdot X \\ \quad \cdot a \cdot X \end{array}$$

, and here there is thus a considerable difference in the way in which one and the other two X's are bonded. Whereas Werner's theory therefore provides no explanation at all of the outstanding tendency of this series of salts to form double salts Ma_6XY_2, this tendency is self-evident according to my theory, and this phenomenon is so unquestionably conspicuous that an explanation of it appears necessary. According to experiments conducted up to the present time XY_2 can be: $(NO_3)(SO_4)$, $(NO_3)(PtCl_6)$, $Cl(PtCl_6)$, $(C_2O_4)_{1/2}(PtCl_6)$, $(PtCl_6)_{1/2}Cl_2$, $(HgCl_3)Cl_2$, $Br(SO_4)$, $Br(PtBr_6)$, $J(SO_4)$, $(PtCl_6)_{1/2}(SO_4)$, $(AuCl_4)(SO_4)$, $(AuBr_4)(SO_4)$, $(SO_4H)(SO_4)$, $(S_2O_8)_{1/2}(SO_4)$, $Cl(SO_4)$, $(J_4O_{11})_{1/2}Cl_2$, and many others.

Also other circumstances point in the same direction. Of course, the hexammine salts are usually very stable compounds which give off ammonia only in exceptional cases. But I have shown that hexamminechromium chloride and, although slowly, hexamminechromium bromide, also on treatment with hydrochloric or hydrobromic acid, exchange $NH_3 \cdot Cl$ and $NH_3 \cdot Br$ for Cl or Br, respectively, with formation of chloro- or bromopentammine chloride.[117] The reason that the process stops at this point and that the other $NH_3 \cdot Cl$ too is not substituted by Cl doubtless is that on the one hand chloropentammine chloride is a compound which is very stable toward acids and that on the other hand praseochromium chloride, which should be formed here, seems completely incapable of existence.[118] In the cobalt series, where the praseo chloride exists, this process can unfortunately not be carried out since the hexamminecobalt chloride is not attacked by hydrochloric acid. However, it turns out that in the cobalt series the other $NH_3 \cdot Cl$ too can be removed since nitropentamminecobalt chloride,

$$\begin{array}{l} \qquad\qquad\qquad\quad \cdot NO_2 \\ Co \cdot a \cdot a \cdot a \cdot a \cdot Cl \\ \qquad\qquad\qquad\quad \cdot a \cdot Cl \end{array}$$

(xanthocobalt chloride), which is completely analogous to chloropentamminechromium chloride, easily exchanges the isolated $NH_3 \cdot Cl$ for NO_2 on suitable treatment with nitrous acid and is

[117] $[[Cr(NH_3)_6]X_3 + HX \rightarrow [Cr(NH_3)_5X]X_2 + NH_4X$ (X = Cl or Br).—G.B.K.]

[118] [Only the violet (*cis*) form is known (L. Grodsenski, *Untersuchungen über Chromiake. Konfigurationsbestimmungsmethoden bei anorganischen Stereoisomeren*, Dissertation, Universität Zürich, 1910).—G.B.K.]

$\cdot NO_2$
thereby converted into croceo chloride, Co·a·a·a·a·Cl.[119] The
$\cdot NO_2$
latter salt, however, is not changed any further, even on prolonged treatment with nitrous acid except insofar as part of it is completely decomposed to cobaltous salt on heating for 24 hours with sodium nitrite and acetic acid; whatever is not completely decomposed is still always present as croceo salt. Werner's theory is unable to give an explanation for such a difference in the behavior of various ammines. Even on prolonged action of ammonium carbonate while
$\cdot OH_2\cdot X$
hot, the aquopentammine salts, Co·a·a·a·a·X, lose the $OH_2\cdot X$
·a·X
and $1NH_3\cdot X$ and are converted into carbonatotetrammine salts, CO_3:Co·a·a·a·a·X, in which the four-membered ammonia group is still intact.[120] On the contrary, the latter is so stable that all compounds which contain their ammonia in only this form can be dissolved in concentrated sulfuric acid, and these solutions, on dropwise addition of concentrated hydrochloric acid, yield almost quantitatively acid praseo sulfate, which contains the four-membered ammonia group still intact.[121] It is regretable that this experiment was not performed with the imidooctammminecobalt salts[79] mentioned above (p. 119).[28] I am certain that no praseo salt will thereby be formed since this series of salts contains only 3- and 2-membered ammonia groups.

Another difference between the two solitary ammonias and the four others in the hexammine salts is shown by the fact that the first two may be exchanged for OH_2, although only indirectly (e.g., by chloropentammine chloride and carbonatotetrammine salt), with formation of aquo and diaquo salts. On the other hand, aquo and diaquo salts which contain OH_2 instead of NH_3 in the four-membered ammonia group (dichro salts[12] and diaquopraseo salts)[122] seem unable to be formed by direct or indirect substitution. For

---

[119] [$[Co(NH_3)_5NO_2]Cl_2 + HNO_2 \rightarrow$ *trans*-$[Co(NH_3)_4(NO_2)_2]Cl + NH_4Cl$.—G.B.K.]

[120] [$[Co(NH_3)_5H_2O]X_3 + (NH_4)_2CO_3 \rightarrow [Co(NH_3)_4CO_3]X + 2NH_4X + NH_3 + H_2O$.—G.B.K.]

[121] [$[Co(NH_3)_4CO_3]X + H_2SO_4 + 2HCl \rightarrow$ *trans*-$[Co(NH_3)_4Cl_2]HSO_4 + HX + CO_2\uparrow + H_2O$ (A. Werner and A. Klien, *Z. anorg. Chem.*, *14*, 28–41 (1897), p. 29).—G.B.K.]

[122] [*trans*-$[Co(NH_3)_2(H_2O)_2Cl_2]X$ (A. Werner, *Z. anorg. Chem.*, *8*, 153–188 (1895), p. 172).—G.B.K.]

their preparation, one must start with compounds which originally contain only 3 or 2 molecules of ammonia. In any case, quite a considerable difference between these two different types of aquo salts is that only the first, i.e., those which contain $OH_2 \cdot X$ bonded to metal, are able to form basic salts with $OH_2 \cdot OH$ instead of $OH_2 \cdot X$. Such basic salts cannot be obtained from those aquo salts which contain OH_2 instead of NH_3 in the four-membered ammonia group.

Furthermore, the oxalopentamminecobalt salts[38] (and similar compounds; see p. 111 above),[28] which, according to my theory, are

$$C_2O_4 \begin{matrix} \cdot NH_3 \\ \cdot Co \cdot a \cdot a \cdot a \cdot a \cdot X, \end{matrix}$$

also show clearly that here one ammonia is bonded in a manner different from the others, since these salts have quite a peculiar tendency to form acid salts of exceptional stability with monobasic acids, e.g., $H \cdot C_2O_4 \cdot Co \begin{matrix} \cdot a \cdot Br \\ \cdot a \cdot a \cdot a \cdot a \cdot Br. \end{matrix}$[123]

Also it is not only the ammonia and water groups which show such a different character. The same is true of the electronegative radicals, which, according to Werner, are bonded at the corners of the octahedron. Thus [A.] Cossa[124] has proved that the remarkable salts which he discovered:

$$Pt \begin{matrix} \cdot NH_3 \cdot Cl \\ \cdot Cl \end{matrix}, ClK \qquad \text{and} \qquad Cl_2 : Pt \begin{matrix} \cdot NH_3 \cdot Cl \\ \cdot Cl \end{matrix}, ClK$$

Potassium platosemiammine chloride[25] and Potassium platinsemiammine chloride[125]

do not behave as they should according to Werner's theory.[125] According to his theory they should be constituted according to the formulas:

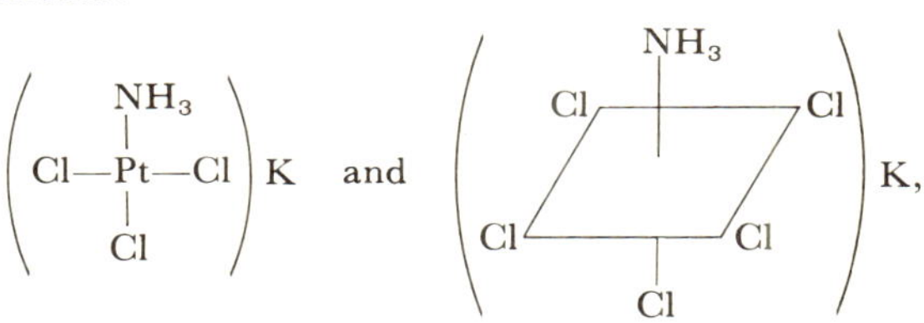

[123] $[[Co(NH_3)_5C_2O_4]Br + HBr \rightarrow [Co(NH_3)_5HC_2O_4]Br_2$ (S. M. Jørgensen, *Z. anorg. Chem.*, *11*, 416–453 (1896), p. 425; A. Werner, *Ann.*, *405*, 212–241 (1914), p. 213).—G.B.K.]

[124] *Z. anorg. Chem.*, *14*, 367 [366–373 (1897)].

[125] [Potassium pentachloroammineplatinate(IV) $(K[Pt(NH_3)Cl_5])$ is known as Cossa's Second Salt.—G.B.K.]

and consequently both salts have all their chlorine atoms bonded in the same manner. On the contrary, Cossa has now found, first that in potassium platosemiammine chloride,[25] for example, only two chlorine atoms are precipitated as silver chloride by silver nitrate at room temperature while the third atom remains bonded to the platinum,[126] but then that the plato salt, on suitable treatment with potassium thiocyanate, yields a thiocyanate $\mathrm{Pt}\begin{matrix}\cdot \mathrm{a} \cdot \mathrm{SNC} \\ \cdot \mathrm{SNC}\end{matrix}$, which contains neither potassium nor chlorine.[127] These reactions definitely show that the original compound is a double salt of $\mathrm{Pt}\begin{matrix}\cdot \mathrm{a} \cdot \mathrm{Cl} \\ \cdot \mathrm{Cl}\end{matrix}$ with KCl and that the three chlorine atoms are in no way bonded in the same manner.

I have also proved that in the diamminecobalt nitrites, which according to Werner's theory are $\left(\mathrm{Co}\begin{matrix}(\mathrm{NH_3})_2 \\ (\mathrm{NO_2})_4\end{matrix}\right)\mathrm{M}$[128] and consequently should have all $4NO_2$ bonded in the same manner, these NO_2 groups, on the contrary, are bonded as pairs in two different ways. Of course, G. Richter[129] has recently tried to disprove this view of mine, but he has not been successful in his reasoning. To be sure, he cites my three reasons for regarding these salts as

$$(\mathrm{NO_2})_2 \cdot \mathrm{Co} \cdot \mathrm{a} \cdot \mathrm{a} \cdot (\mathrm{NO_2})_2 \cdot \mathrm{M}, \quad \text{viz.,}$$

---

[126] [The reaction occurs in two steps, of which the second is uncertain:

$$\mathrm{K[Pt(NH_3)Cl_3] + AgNO_3 \rightarrow Ag[Pt(NH_3)Cl_3] + KNO_3}$$
$$\mathrm{Ag[Pt(NH_3)Cl_3] + AgNO_3 \rightarrow Ag[Pt(NH_3)Cl_2NO_3] + AgCl \downarrow .}$$

—G.B.K.]

[127] [The yellow-brown amorphous precipitate has not been well characterized. It may have the constitution

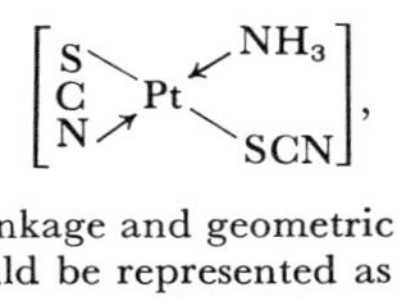

with the possibility of both linkage and geometric isomerism. The thiocyanate or isothiocyanate group should be represented as —SCN or —NCS but not as —SNC, as Jørgensen has shown it.—G.B.K.]

[128] [These compounds, of which the ammonium salt is known as Erdmann's salt, were first discovered in 1866 (O. L. Erdmann, *J. prakt. Chem.*, *97*, 406, 409 (1866)) and exist in the *trans* form only (Y. Shibata, *J. Coll. Sci. Imp. Univ. Tokyo*, *37*, Art. 8 (1916), pp. 18 and 20).—G.B.K.]

[129] [A. Werner and G. Richter,] *Z. anorg. Chem.*, *15*, 243[–277 (1897)].

1. that ammonia splits off $2NH_4 \cdot NO_2$ from ammonium diamminenitrite with formation of a flavo salt;[130,131]

2. that hydrochloric acid splits off $2HNO_2$ from the diamminenitrite radical in croceocobalt diamminenitrite;[132,133]

3. that oxalic acid splits off $2HNO_2$ from ammonium diamminenitrite, while C_2O_4 replaces the two NO_2 groups with formation of ammonium dinitrodiamminecobaltoxalate.[134,135]

G. Richter now raises various objections to the first of these reasons. But *these have significance only* in the case in which ammonia is allowed to act upon ammonium diamminenitrite *without addition of an ammonium salt.* Here the process proceeds differently under slightly altered conditions, and in this case the proof is of course not satisfactory. On the other hand, he seems completely to have overlooked the fact that I later showed[136] that upon addition of an ammonium salt, the process,

$$NH_4 \cdot (NO_2)_4(NH_3)_2 \cdot Co + 2NH_3 + NH_4 \cdot NO_3$$
Ammonium diamminenitrite
$$= (NO_2)_2 \cdot Co \cdot (NH_3)_4 \cdot NO_3 + 2NH_4 \cdot NO_2,$$ [137]
Flavo nitrate

which proceeds quite smoothly, occurs, and in this form the proof is supposedly unquestionable.

G. Richter has offered no [valid] objections to the second of my reasons. For the fact that the croceo radical is affected by hydrochloric acid, as croceo salts usually are, which is the only point with which he finds fault, does not at all concern the question at hand. He makes no comment at all about the third of my reasons. Nevertheless, he concludes:[138] "The fact that under certain circumstances

---

[130] [S. M. Jørgensen,] *ibid.*, *5*, 156 and 182 [147–196 (1893)].

[131] [2{NH_4 *trans*-[$Co(NH_3)_2(NO_2)_4$]} + $2NH_3$ → *cis*-[$Co(NH_3)_4(NO_2)_2$]*trans*-[$Co(NH_3)_2(NO_2)_4$] + $2NH_4NO_2$.—G.B.K.]

[132] [S. M. Jørgensen,] *Z. anorg. Chem.*, *13*, 184 ff. [172–190 (1896)].

[133] [*trans*-[$Co(NH_3)_4(NO_2)_2$]*trans*-[$Co(NH_3)_2(NO_2)_4$] (croceo double salt) + 3HCl → *trans*-[$Co(NH_3)_4(NO_2)Cl$]*trans*-[$Co(NH_3)_2(NO_2)_2Cl_2$] + $3HNO_2$.—G.B.K.]

[134] [S. M. Jørgensen,] *Z. anorg. Chem.*, *11*, 418 and 440 [416–453 (1896)].

[135] [NH_4 *trans*-[$Co(NH_3)_2(NO_2)_4$] + $H_2C_2O_4$ → NH_4 *trans*-[$Co(NH_3)_2(C_2O_4)(NO_2)_2$] + $2HNO_2$.—G.B.K.]

[136] *Z. anorg. Chem.*, *7*, 314 [289–330 (1894)].

[137] [NH_4 *trans*-[$Co(NH_3)_2(NO_2)_4$] + $2NH_3$ + NH_4NO_3 → *cis*-[$Co(NH_3)_4(NO_2)_2$]NO_3 + $2NH_4NO_2$.—G.B.K.]

[138] [A. Werner and G. Richter,] *ibid.*, *15*, 247 [243–277 (1897)].

NO_2 groups are removed from diamminecobaltnitrites in pairs thus does not seem sufficient to deduce a different type of bonding for each two of these groups in the molecule." The conclusion is obviously completely untenable. For he cites one of my reasons incompletely, and he does not mention the other two at all. The fact that the $4NO_2$ groups in the diamminenitrites are bonded differently is irrefutable, and since in all the three processes cited the transformation product contains $2NO_2$ groups bonded to cobalt, this must also be the case in the original compound. If the two others are bonded in a different manner, it can only be to ammonia. The fact that a compound which is analogous to ammonium diamminecobaltnitrite[128] in several respects, viz., ammonium diamminechromiumthiocyanate,[139] behaves in another manner in quite different processes cannot possibly weaken the experimental results which were furnished by the nitrite salt itself.[140]

---

[139] [Ammonium tetrathiocyanatodiamminechromate(III), $NH_4[Cr(NH_3)_2(SCN)_4]$, is known as Reinecke's Salt (A. Reinecke, *Ann.*, *126*, 113–118 (1863))and is known only in the *cis* form.—G.B.K.]

[140] Here too, incidentally, G. Richter has not been fortunate in his reasoning. Inasmuch as I had often emphasized that, as a pure nitrite, triamminecobalt nitrite[111] does not seem suited to serve as evidence for Werner's fundamental principle, since the constitution of so many nitrites can be, and in fact is, interpreted in different ways, he wants to test the case by investigating a compound which contains no nitro groups at all, and he now chooses a thiocyanate, viz., Reinecke's ammonium diamminechromiumthiocyanate,[139] and some compounds prepared from it. But what is true of compounds of the radical NO_2 is of course true to an even greater extent of those of the radical CNS.[127] Just as NO_2 appears sometimes as $\cdot\overset{V}{N}O_2$, sometimes as $\cdot O\cdot\overset{III}{N}:O$, so does CNS appear sometimes as $\cdot S\cdot\overset{V}{N}\cdot\vdots C$,[127] sometimes as $\overset{II}{N}:C:S$,[127] and many double salts of nitrous and thiocyanic acids are, to judge from all the evidence, again to be interpreted in a somewhat different manner than the single salts. In addition to this, however, it happens that we know for certain from the existence and behavior of thiocyanic acid and isothiocyanic acid, etc., that three thiocyanate radicals bonded to each other can combine to form a trivalent unit and to be sure in at least two different ways. From this special tendency it could then be explained very well also that Reinecke's Salt, which contains 4 thiocyanate groups, splits off only one and that the resulting compound does not contain ions. Exactly what G. Richter intended to disprove, viz., that $(NO_2)_3$ can occur as a trivalent complex, is of course quite common for $(CNS)_3$. And if he detects an isomerism (the two compounds $Cr(NH_3)_2\cdot(OH_2)_2(CNS)_3$)[141] which he considers to be incompatible with my theory, then this proves just the opposite, that these compounds are actually incompatible with Werner's theory, because an octahedron has no more than 6 corners, but here are 7 groups of

In triamminecobaltnitrite[111] too, the type of bonding of all 3 nitro groups should be exactly the same according to Werner's formula $\left[Co \begin{matrix}(NH_3)_3\\(NO_2)_3\end{matrix}\right]$. But here too it seems to me that the experimental results definitely favor the bonding of one nitro group in a special manner. Certainly, by heating of the salt with concentrated hydrochloric acid, 3 nitro groups are replaced by chlorine with formation of dichro chloride.[142] But in less vigorous reactions only one nitro group is eliminated. Thus triamminenitrite yields dinitrotriamminechloride, $(NO_2)_2 \cdot Co \cdot (NH_3)_3 \cdot Cl$ with cold half-concentrated [1:1] hydrochloric acid,[143] and conversely, dinitrotriamminechloride smoothly yields sodium chloride and triamminenitrite with sodium nitrite.[144] Likewise, triamminenitrite, on being warmed briefly with ammonia and ammonium nitrate, yields flavo nitrate by replacing one nitro group of the triamminenitrite with $NH_3 \cdot NO_3$.[145]

Werner's theory finds itself in a very peculiar dilemma with respect to the croceo salts. He denies that the isomerism of this series of salts with the flavo salts is caused by a difference in the NO_2

---

atoms which are all supposed to be bonded to the chromium atom. I see no reason to go any further into the explanation of the two different aquo salts. They require a more thorough investigation before anything at all can be said about their constitution. Indeed, it is not even certain that they are diaquo compounds at all. At least, their behavior toward metal salts seems to point in an entirely different direction.

[141] [The compound $[Cr(NH_3)_2(H_2O)_2(SCN)_3]$ of apparent coordination number seven (Jørgensen's paper mistakenly shows this with *Co* instead of *Cr*) was first prepared by O. Nordenskjöld by oxidation of Reinecke's Salt[139] with H_2O_2 (*Z. anorg. Chem.*, *1*, 126–143 (1892), p. 137). Werner and Richter[138] described this salt as an α-compound (p. 268) and claimed to have also prepared a β-form by oxidation of Reinecke's Salt with HNO_3 (p. 273), an isomerism that has not been definitely established (P. Pfeiffer, *Ber.*, *37*, 4255–4290 (1904), p. 4256 footnote). The compound should be written $[Cr(NH_3)_2(H_2O)(SCN)_3] \cdot H_2O$.—G.B.K.]

[142] [*mer*-$[Co(NH_3)_3(NO_2)_3] + 3HCl + H_2O \rightarrow$ *trans*-$[Co(NH_3)_3(H_2O)Cl_2]Cl + 3HNO_2$.—G.B.K.]

[143] [*mer*-$[Co(NH_3)_3(NO_2)_3] + HCl \rightarrow [Co(NH_3)_3(NO_2)_2Cl] + HNO_2$.—G.B.K.]

[144] [$[Co(NH_3)_3(NO_2)_2Cl] + NaNO_2 \rightarrow$ *mer*-$[Co(NH_3)_3(NO_2)_3] + NaCl$.—G.B.K.]

[145] [*mer*-$[Co(NH_3)_3(NO_2)_3] + NH_3 + NH_4NO_3 \rightarrow$ *cis*-$[Co(NH_3)_4(NO_2)_2]NO_3 + NH_4NO_2$.—G.B.K.]

groups; on the contrary, he assumes a position isomerism (*Orts-isomerie*):[146]

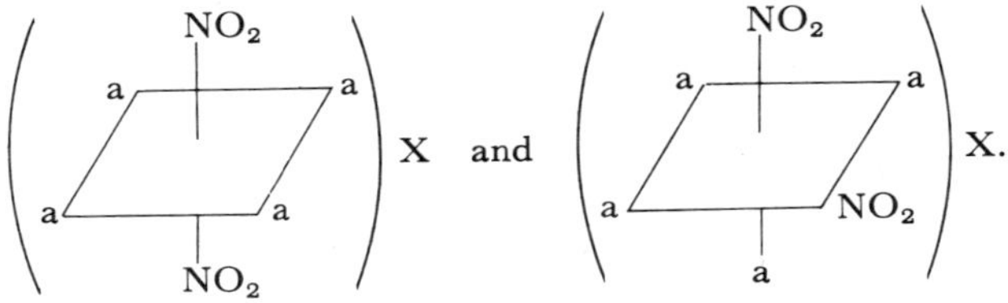

Croceo salt Flavo salt

According to him, the croceo salts belong to the praseo series, the flavo salts to the ammonia-violeo series,[10] which, incidentally, is completely unknown.[6] The fact that praseo salts are transformed into croceo salts on being warmed with nitrous acid[147] agrees with this. With this assumption, however, the following remains unintelligible: (1) that the two NO_2 groups in the croceo salts, although identical and, as coordinated to the cobalt atom, bonded in exactly the same manner, behave quite differently toward acids since one is thereby very easily eliminated, while the other is not;[148] (2) that even on being boiled with hydrochloric acid, no praseo salt is formed, and (3) that, on the other hand, the flavo salts, on being warmed with hydrochloric acid, form praseo salt in considerable amounts.[149,150] One could think that Werner, without otherwise changing his system, could, on this point, probably assume with me that in the croceo salts one NO_2 is a nitro group and the other a nitrito group, while the flavo salts contain a dinitrito group.[151] In this way, the above difficulties would certainly be removed, but at the same time a position isomerism[146] between the croceo and flavo salts would become completely superfluous, the only reason for the assumption of ammonia-violeo salts would disappear, among all cobalt-,[6] chromium-,[152] rhodium-, and iridium-ammonia salts not

---

[146] [That is, stereoisomerism (*Raumisomerie*).—G.B.K.]

[147] [*trans*-$[Co(NH_3)_4Cl_2]X + 2HNO_2 \rightarrow$ *trans*-$[Co(NH_3)_4(NO_2)_2]X + 2HCl$.—G.B.K.]

[148] See Paper Nine on the *trans*-effect in Kauffman, *Classics*, Part III.

[149] [S. M. Jørgensen,] *Z. anorg. Chem.*, *14*, 408 [404–422 (1897)].

[150] [*cis*-$[Co(NH_3)_4(NO_2)_2]X + 2HCl \rightarrow$ *trans*-$[Co(NH_3)_4Cl_2]X + 2HNO_2$.—G.B.K.]

[151] [Jørgensen considered these isomers to be *structural isomers*, while Werner considered them to be *stereoisomers*.—G.B.K.]

[152] [Compounds of the type *cis*-$[Cr(NH_3)_4(Cl_2 \text{ or } Br_2)]X$ have been isolated.[118]—G.B.K.]

even one would be found, which would have an electronegative radical in one corner of the square, the possibility of such compounds would thus become improbable to a high degree, and the state of affairs would thereby become still more questionable for Werner's theory.

Because of the reason just cited the ammonia-violeo salts have an extraordinary importance for Werner's theory,[6] and therefore he tries to preserve their existence even if he has to sacrifice simplicity. I have already mentioned above (p. 105) that he formerly counted chloroaquotetrammine chloride among the praseo series, whereas now, after I had showed that it yields flavo chloride with nitrous acid, he assumes that it belongs to the violeo series.[153] Although it is formed from praseo chloride under exactly the same conditions (warming with very dilute acid) and even more easily than aquopentammine chloride is formed from chloropentammine chloride and although the process is the same in both cases, viz., the substitution of $OH_2 \cdot Cl$ for Cl:

$$\text{Chloropentammine chloride: } Co \begin{array}{l} \cdot Cl \\ \cdot a_4 \cdot Cl \\ \cdot a \cdot Cl \end{array} + H_2O$$

$$= Co \begin{array}{l} \cdot OH_2 \cdot Cl \\ \cdot a_4 \cdot Cl \\ \cdot a \cdot Cl \end{array} \quad \text{: Aquopentammine chloride,}^{154}$$

$$\text{Praseo chloride: } Co \begin{array}{l} \cdot Cl \\ \cdot a_4 \cdot Cl \\ \cdot Cl \end{array} + H_2O$$

$$= Co \begin{array}{l} \cdot Cl \\ \cdot a_4 \cdot Cl \\ \cdot OH_2 \cdot Cl \end{array} \quad \text{: Chloroaquotetrammine chloride,}^{155}$$

now[156] the salt is supposed to be formed according to the scheme [at the head of the next page].

In order to maintain the existence of ammonia-violeo salts, Werner must therefore assume in this simple process the existence

---

[153] [*cis*-$[Co(NH_3)_4(H_2O)Cl]Cl_2 + 2HNO_2 \rightarrow$ *cis*-$[Co(NH_3)_4(NO_2)_2]Cl + 2HCl + H_2O$. The chloroaquo compound is indeed *cis* (violeo) not *trans* (praseo).—G.B.K.]

[154] [$[Co(NH_3)_5Cl]Cl_2 + H_2O \rightarrow [Co(NH_3)_5H_2O]Cl_3$.—G.B.K.]

[155] [*trans*-$[Co(NH_3)_4Cl_2]Cl + H_2O \rightarrow$ *cis*-$[Co(NH_3)_4(H_2O)Cl]Cl_2$.—G.B.K.]

[156] [A. Werner,] *Z. anorg. Chem.*, *15*, 148 [143–172 (1897)].

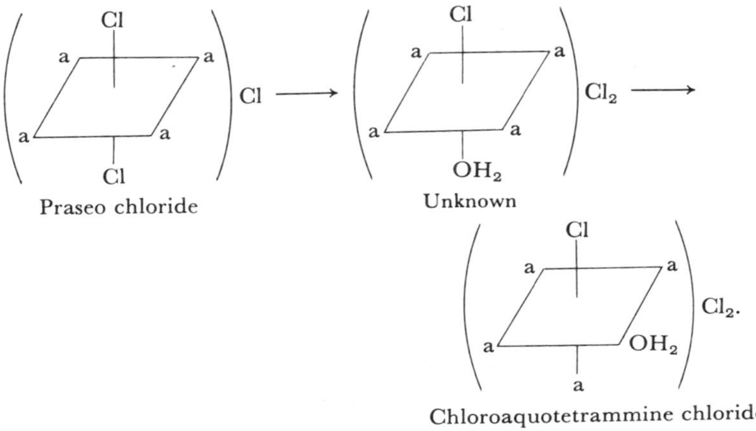

of an unknown compound and the rearrangement of an NH_3 and an OH_2 group, which is contrary to all other evidence.[157] And such unknown compounds and such rearrangements must always be assumed whenever praseo salts are transformed into other series

[157] In a completely analogous case the assumed rearrangement does not in fact take place, and the "characteristic of the water molecule of favoring the position *ortho* to the negative radical" ([A. Werner,] *Z. anorg. Chem.*, *15*, 148 [143–172 (1897)]) does not in fact appear. Chloronitrotetrammine chloride must belong to the praseo series because it is so exceptionally easily formed from croceo salts and is so easily converted to aquoxantho chloride. We therefore should have:

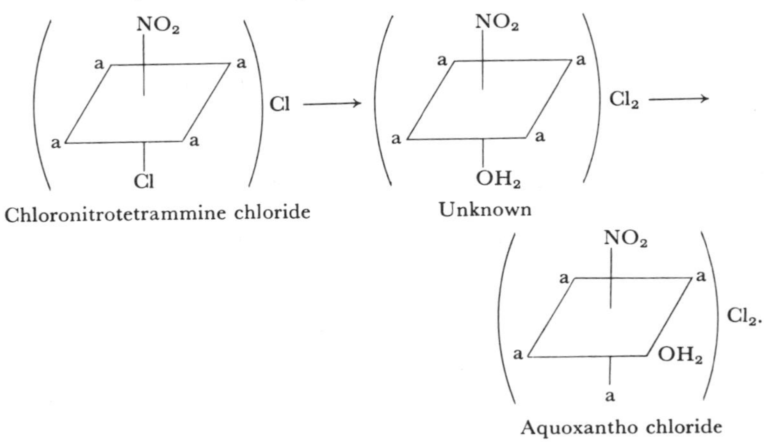

But the resulting aquoxantho chloride yields no flavo salt with sodium nitrite and acid, but instead the calculated quantity of croceo salt.[158] A rearrangement can therefore not have taken place.

of salts, e.g., if they are transformed with ammonium carbonate into carbonatotetrammine salts,[159] with oxalic acid into oxalotetrammine salts,[160] or with dilute sodium hydroxide into basic diaquotetrammine salts, for Werner considers all these series of salts as violeo salts. Conversely, a rearrangement in the opposite direction and the formation of unknown compounds as transition compounds must take place whenever flavo salts form praseo salts with hydrochloric acid or whenever all other tetrammine salts form acid praseo sulfate on being dissolved in concentrated sulfuric acid followed by addition of hydrochloric acid. I think that one must admit that the older theory, which requires no rearrangement and no unknown compounds in all these processes, has at least the advantage of simplicity.[161]

And despite all these numerous rearrangements Werner is unable to prove the existence of ammonia-violeo salts.[6] He *finds* that croceo salts are formed from praseo salts and therefore belong to the same series. But he *assumes* that the flavo salts, which are isomeric with the croceo salts, can be formed from an *unknown* series of violeo salts which is isomeric with the praseo salts, and from this *assumption* he further concludes that, since the flavo salts are formed from chloroaquotetrammine salts, the latter must belong to the ammonia-violeo series. This is a conclusion which rests on an assumption, but not a proof. In fact, no violeo chloride isomeric with praseo chloride[6] and no praseo compounds isomeric with chloroaquotetrammine chloride[162] or with the other salts considered as belonging to the violeo series are known.

The case just cited is by no means the only one in which Werner's

[158] [The chloronitro compound is indeed now known to belong to the *trans* (praseo) series. Aquoxantho salts, however, are generally believed to be *trans* compounds also and thus correspond to the postulated unknown intermediate (J. Meyer and O. Rampoldt, *Z. anorg. Chem.*, *214*, 1–15 (1933)), and no rearrangement is needed to explain their formation. *Cis*-$[Co(NH_3)_4(H_2O)NO_2]X_2$ compounds have also been described (R. G. Yalman and T. Kuwana, *J. Phys. Chem.*, *59*, 298–300 (1955); H. J. S. King and S. N. Mistry, *J. Chem. Soc.*, *1957*, 2402–2404).—G.B.K.]

[159] [S. M. Jørgensen,] *Z. anorg. Chem.*, *14*, 409 [404–422 (1897)].

[160] [S. M. Jørgensen,] *ibid.*, *11*, 430 [416–453 (1896)].

[161] [Werner was later able to explain many such apparent inconsistencies by a theory of rearrangements (*Ann.*, *386*, 1–272 (1912)).—G.B.K.]

[162] [The preparation of *trans*-$[Co(NH_3)_4(H_2O)Cl]Cl_2$ has been reported (M. G. ter Horst (*Rec. trav. chim.*, *54*, 257–274 (1935), p. 269; R. Tsuchida, *Bull. Chem. Soc. Japan*, *11*, 721–734 (1936), p. 723).—G.B.K.]

octahedral formulas require types of isomerism which have never been observed. Although he has abandoned the view that the electronegative ions lie at the principal points of his octahedron and could thereby cause isomerisms, yet there remains a not small number of cases where, according to his theory, types of isomerism which are not assumed by the older theory and which are completely unknown must be capable of existence. Thus, according to Werner's theory:

1. All compounds $(Ma_4X_2)Y$, all compounds $(Ma_4\overset{II}{R})Y$,[163] and all compounds $(MX_4a_2)R$ should occur in two modifications, therefore, ammonia-praseo salts, dinitrotetrammine salts, carbonatotetrammine salts, oxalotetrammine salts, sulfitotetrammine salts,[164] and diamminenitrites.[165] Of the dinitrotetrammine salts only, two isomeric series are known,[10] whose isomerism, however, probably must be explained in a different manner,[151] as will soon be discussed in more detail.

2. The compounds $(Ma_4b_2)X_3$ should exist in two forms. Only the diaquotetrammine salts are known.[166]

3. Three isomeric forms of compounds $(Ma_3hCl_2)X$ and $(Ma_3(NO_2)_2)Cl$[167] should exist. Of the first, there is known with certainty (See p. 105)[28] only the dichro salts[12,14] and only *one* dinitrotriammine chloride.[168] The triammine nitrite should be capable of existing in two forms. According to my thorough investiga-

---

163 [These compounds exist in only one modification.—G.B.K.]

164 Completely apart from types of isomerism which can be caused by the different function of the sulfurous acid as $HO\cdot\overset{IV}{S}O\cdot OH$ and as $H\cdot\overset{VI}{S}O_2\cdot OH$, since these are possible also according to the older theory.

165 [$[Co(NH_3)_4X_2]Y$, $[Co(NH_3)_4(NO_2)_2]X$, $[Co(NH_3)_4CO_3]X$, $[Co(NH_3)_4$-$C_2O_4]X$, $[Co(NH_3)_4SO_3]X$, $M[Co(NH_3)_2(NO_2)_4]$, respectively.—G.B.K.]

166 [Although the corresponding ethylenediamine compounds, $[Coen_2(H_2O)_2]X_3$, have long been known in two modifications, until relatively recently the so-called tetrammineroseo salts, $[Co(NH_3)_4(H_2O)_2]X_3$, have been known in only the *cis* configuration. The *trans*-sulfate (M. G. ter Horst, *Rec. trav. chim.*, *54*, 257–274 (1935), p. 261) and *trans*-nitrate (R. G. Yalman and T. Kuwana, *J. Phys. Chem.*, *59*, 298–300 (1955)) have now been isolated.—G.B.K.]

167 [Jørgensen insists that $[M(NH_3)_3(NO_2)_2Cl]$ compounds are electrolytes because of the abnormally high conductivities due to aquation.[99]—G.B.K.]

168 [While the general chemical behavior of $[Co(NH_3)_3(NO_2)_2Cl]$ is in accord with the *cis*-peripheral structure (I. I. Chernyaev and A. Ganiev, *Zhur. Neorg. Khim.*, *2*, 765–771 (1957); *J. Inorg. Chem. U.S.S.R.*, *2*, No. 4, 85–97 (1957), p. 87), the *trans*-peripheral configuration has been reported on the basis of three-dimensional Fourier analysis (Y. Tanito, Y. Saito, and H. Kuroya, *Bull. Chem. Soc. Japan*, *26*, 420–424 (1953)).—G.B.K.]

tions[169] on this point, I certainly believe that only one form occurs.[111]

4. All compounds $(Ma_4hX)Y_2$ and $(Ma_4XY)Z$ should occur in two isomeric modifications, such as the chloroaquotetrammine salts,[153] the aquoxantho salts,[158] and chloronitrotetrammine chloride.[158] None of these is known in more than one modification.[158]

5. Five isomeric series of dichlorodiaquodiammine salts[170] and also five isomeric series of dinitrodiammine oxalates[171] should be found. Only one of each is known.

On the other hand, according to Werner's system now, only one series of nitropentammine salts is possible, except that he assumes with me a difference between nitro and nitrito compounds. But two nitropentammines are known, viz., the xantho and isoxantho salts.[172,173] Now the way in which Werner expresses himself concerning them,[174] viz., that "one is involuntarily forced to make the assumption of seeing analogous series of compounds in these two series of salts," is certainly not clear, but it can at least be interpreted to mean that he too regards them as analogous, with only a difference in the NO_2 groups. I find it highly probable that such a difference

---

[169] *Z. anorg. Chem.*, *13*, 172[–190 (1896)].

[170] [$[Co(NH_3)_2(H_2O)_2Cl_2]Cl$ is known in two modifications (green and blue), which, according to Werner (*Z. anorg. Chem.*, *15*, 143–172 (1897), p. 152) are probably stereoisomers.—G.B.K.]

[171] [If the two ammonia molecules in compounds $M[Co(NH_3)_2(C_2O_4)(NO_2)_2]$ are in the *trans* position to each other, the salts should exist in one nonresolvable form. On the other hand, if they are *cis* to each other, two forms are possible—an inactive, nonresolvable form if the NO_2 groups are *trans* to each other (*trans-cis*) and a racemic, resolvable form if the NO_2 groups are *cis* to each other (*cis-cis*). By fractional crystallization of the barium salt, W. Thomas (*J. Chem. Soc.*, *123*, 617–619 (1923)) obtained a monoclinic and a rhombic form, the latter of which he was able to resolve. He also obtained the NH_4^+ and K^+ salts in monoclinic forms. He therefore believed that the monoclinic salts were *trans-cis* isomers and that the rhombic salts were *cis-cis* isomers. E. H. Riesenfeld and R. Klement (*Z. anorg. Chem.*, *124*, 1–21 (1922), pp. 5, 14), however, believe that the NH_3 molecules are in the *trans* positions in this series.—G.B.K.]

[172] [S. M. Jørgensen,] *Z. anorg. Chem.*, *5*, 169, 172 ff. [147–196 (1894)].

[173] [Jørgensen felt that Werner was being inconsistent in agreeing with him that the xantho and isoxantho salts were structural isomers yet insisting that the flavo and croceo salts were stereoisomers. The formulas that Werner assigned to these compounds—the ones accepted today—are as follows: xantho, $[Co(NH_3)_5NO_2]X_2$; isoxantho, $[Co(NH_3)_5ONO]X_2$; flavo, *cis*-$[Co(NH_3)_4(NO_2)_2]X$; croceo, *trans*-$[Co(NH_3)_4(NO_2)_2]X$.—G.B.K.]

[174] [A. Werner and A. Miolati,] *Z. physik. Chem.*, *14*, 513 [506–521 (1894)]. For a discussion and English translation see Kauffman, *Classics*, pp. 117–139 (p. 129). —G.B.K.]

actually exists because of the following experiments. If freshly prepared isoxantho chloride is warmed slightly for a moment with normal sulfuric acid, it dissolves in the form of ordinary roseo sulfate, with evolution of nitrous acid.[175] With H_2PtCl_6 and $Na_4P_2O_7$, the solution gives the usual roseo sulfate reactions, and the precipitations are almost complete. Also, fresh isoxantho chloride dissolves with evolution of nitrous acid when it is warmed slightly for a very short time with very dilute hydrochloric acid.[176] The solution contains roseo chloride and is precipitated in the usual manner by potassium ferricyanide and by cooling with concentrated hydrochloric acid. The latter precipitate is roseo chloride and is transformed completely into chloropentammine chloride when heated with the liquid for a few minutes on the water bath. According to this it can scarcely be doubted that isoxantho salts are nitritopentammine salts, and it thereby becomes extremely probable that the isomeric but much more stable xantho salts are to be viewed as nitropentammine salts.[173] But if this is the case, I do not understand why the identical difference between the two NO_2 groups in the croceo salts should not have the same cause.[173] Croceo chloride too, on being boiled with normal hydrochloric acid, evolves one of the NO_2 groups as nitrous acid and is transformed into aquoxantho chloride.[177] Even on heating 1 g. of croceo chloride with 4 ccm. of normal hydrochloric acid, 6 ccm. of ammonium chloride solution (1:15), and 10 ccm. of water just until the salt is dissolved, nitrogen is evolved in copious amounts, the solution becomes dark red and almost solidifies on cooling because of the crystallization of aquoxantho chloride. Neither crystals nor mother liquor contain croceo chloride in detectable amounts. The solution of the crystals as well as the mother liquor are almost completely precipitated by ammonium oxalate. (Croceo salts give no precipitate with this reagent.) With dilute nitric acid both give not needles of croceo nitrate,[178] but on standing aquoxantho nitrate in small octahedra. From both, ammonium sulfate (1:5) precipitates not quadratic plates of

---

[175] [$2[Co(NH_3)_5ONO]Cl_2 + 3H_2SO_4 + 2H_2O \rightarrow [Co(NH_3)_5H_2O]_2(SO_4)_3 + 2HNO_2 + 4HCl$.—G.B.K.]

[176] [$[Co(NH_3)_5ONO]Cl_2 + HCl + H_2O \rightarrow [Co(NH_3)_5H_2O]Cl_3 + HNO_2$.—G.B.K.]

[177] [*trans*-$[Co(NH_3)_4(NO_2)_2]Cl + HCl + H_2O \rightarrow [Co(NH_3)_4(H_2O)NO_2]Cl_2 + HNO_2$ (a little *trans*-$[Co(NH_3)_4Cl_2]Cl$ (praseo chloride) is also formed).—G.B.K.]

[178] See [S. M. Jørgensen,] *Z. anorg. Chem.*, *17*, 471 and 472 ff. [455–479 (1898)].

croceo sulfate but rather rhomboidal plates of 62°, i.e., aquoxantho sulfate.

Therefore it is supposedly indubitable that one NO_2 in the croceo salts is present as a nitrito group.[173] But in this case the flavo salts, which do not show such a difference between the $2NO_2$ groups since the latter are *simultaneously* eliminated by hydrochloric acid even at room temperature and even more easily on heating,[179] are naturally to be viewed as dinitrito- or perhaps as dinitrotetrammine salts.[173]

In these relations too, viz., concerning the number of isomers and the explanation of the cases of isomerism of the nitropentammine and dinitrotetrammine salts, the older theory seems to me to agree better with the facts than Werner's theory.

But if the latter therefore requires numerous compounds which have never been observed and whose existence is actually supported only by Werner's theory, there are on the other hand numerous compounds which actually exist and which this theory is unable to explain. I have discussed above the difficulties which are presented to Werner's theory by dinitrotriammine chloride, the sulfato-, oxalo-, and sulfitopentammine salts (or tetrammine salts), Hofmann's thiocarbonate salts, the oxy- and anhydrooxycobaltammine salts, the different imidooctammine salts, the dinitrodiammine oxalates, the diammine nitrites, and the croceo and flavo salts. But furthermore, there are still other metal-ammonia salts of the series to be considered here, of which I have absolutely no idea how they can at all be explained according to his theory.[180] Thus the acid hexammine and aquopentammine nitrates[181]

$Ma_6(NO_3)_3, HNO_3$[182] where M = Co, Cr, Rh,
$Ma_5h(NO_3)_3, HNO_3$[183] ,, ,, = Co, Cr, Rh;

---

[179] [S. M. Jørgensen,] *Z. anorg. Chem.*, *14*, 408 [404–422 (1897)].

[180] [The modern formulas for the following compounds are:

$[M(NH_3)_6](NO_3)_3 \cdot HNO_3$,
$[M(NH_3)_5H_2O](NO_3)_3 \cdot HNO_3$,
trans-$[Coen_2Br_2]Br \cdot HBr \cdot 2H_2O$,
trans-$[Coen_2Cl_2]Cl \cdot HCl \cdot 2H_2O$,
trans-$[Rh(C_5H_5N)_4Cl_2]Cl \cdot HCl \cdot 2H_2O$,
cis-$[Co(NH_3)_4(NO_2)_2]NO_3 \cdot HNO_3$,
cis-$[Co(NH_3)_4(NO_2)_2]NO_3 \cdot KNO_3$,
and *trans*-$[Co(NH_3)_4(NO_2)Cl]$*trans*-$[Co(NH_3)_2(NO_2)_2Cl_2]$,
respectively.—G.B.K.]

[181] [S. M. Jørgensen,] *J. prakt. Chem.* [2], *44*, 63[–66 (1891)].

[182] [The original paper erroneously reads "Ma_6NO_3, HNO_3."—G.B.K.]

[183] [The original paper erroneously reads "Ma_5hNO_3, HNO_3."—G.B.K.]

thus the acid bromo- and chloro-ethylenediaminepraseo salts and a corresponding rhodium pyridine salt:[184]

$$Br_2 \cdot Co \cdot en_2 \cdot Br, HBr, 2H_2O \quad (en = \text{ethylenediamine}),$$
$$Cl_2 \cdot Co \cdot en_2 \cdot Cl, HCl, 2H_2O,$$
$$Cl_2 \cdot Rh \cdot p_4 \cdot Cl, HCl, 2H_2O \quad (p = \text{pyridine});$$

thus the acid flavo nitrate[185] and a corresponding potassium salt, which I described recently:[186]

$$(NO_2)_2Co \cdot a_4 \cdot NO_3, HNO_3,$$
$$(NO_2)_2 \cdot Co \cdot a_4 \cdot NO_3, KNO_3;$$

and thus chloronitrotetrammine dinitrodiamminechloride:[187]

$$\begin{matrix} NO_2 \\ Cl \end{matrix} > Co \cdot a_4 \cdot Cl_2 \cdot a_2 \cdot Co \cdot (NO_2)_2.$$

In particular, the compound $[(NO_2)_2 \cdot Co \cdot a_2]_2C_2O_4$[188] also does not seem at all consistent with Werner's theory, whether it is regarded as dinitrodiammine oxalate or as dinitrodiammine dinitrodiammine oxalate.[189] Of course, he has tried to force the salt into his system,[190] but by assuming that C_2O_4 occupies *one* coordination position in the oxalopentammine salts, *two* coordination positions in the oxalotetrammine salts, and *four* coordination positions in dinitrodiammine oxalate.[50] It is clear that with such assumptions anything possible can be proved. The same is true of the way in which G. Richter[191] tries to explain the $Cra_2(OH_2)_2(CSN)_3$ prepared by O. Nordenskjöld and an isomeric compound prepared by Richter himself.[141] That is, in order to fit them into Werner's system, he simply assumes, without mentioning any reason for it,

---

[184] [S. M. Jørgensen,] *J. prakt. Chem.*, *39*, 24, 25 [1–26 (1889)]; *41*, 444 [440–459 (1890)].
[185] [S. M. Jørgensen,] *Z. anorg. Chem.*, *5*, 166 [147–196 (1894)].
[186] *Ibid.*, *17*, 473 [455–479 (1898)].
[187] [S. M. Jørgensen,] *ibid.*, *13*, 183 ff. [172–190 (1896)].
[188] [S. M. Jørgensen,] *ibid.*, *11*, 451 [416–453 (1896)].
[189] [That is, as $[Co(NH_3)_2(NO_2)_2]_2C_2O_4$ or as $[Co(NH_3)_2(NO_2)_2][Co(NH_3)_2(C_2O_4)(NO_2)_2]$. In either case, the cobalt atom in the cation has an apparent coordination number of only four. This unusual compound is listed in *Gmelins Handbuch der Anorganischen Chemie*, 8th ed., System Nr. 58, Teil B, Verlag Chemie G.M.B.H., Berlin, 1930, p. 330, as a compound of type $[(CoA_2)_2Y^{II}X_4{}^{I}]$. It is assigned the name "oxalatobisdinitrodiamminedicobalt(III)" with the formula "$[(Co(NH_3)_2)_2(C_2O_4)(NO_2)_4]$?."—G.B.K.]
[190] [A. Werner, *Z. anorg. Chem.*,] *14*, 26 [21–27 (1897)].
[191] [A. Werner and G. Richter,] *ibid.*, *15*, 269 [243–277 (1897)].

that 2 molecules of water here act as one. Such completely arbitrary assumptions are beyond serious criticism, and how G. Richter can maintain such an assertion at all is almost incomprehensible since [O.] Nordenskjöld has already refuted it[192] experimentally for his compound by proving that one of the two water molecules can be replaced by piperidine and ethylenediamine and both water molecules by aniline. For although I am used to many strange conclusions in the controversy against the older theory of metal-ammonia salts, yet I can scarcely believe that one wants to assert that also 2 molecules of aniline or 1 molecule of water + 1 molecule of piperidine act "bimolecularly" and can replace 1 molecule of NH_3 or 1 molecule of water.

But to these compounds, which present Werner's theory with serious and in most cases even scarcely surmountable difficulties, are now added the numerous salts which contain water. I do not wish to dwell upon the fact that a great number of the most common salts (potassium chloride, potassium nitrate, sodium carbonate, Glauber's salt [$Na_2SO_4 \cdot 10H_2O$], ordinary sodium phosphate, sodium pyrophosphate, borax, sodium thiosulfate, copper sulfate pentahydrate,[193] ferrous sulfate heptahydrate,[194] the alums, etc.) either do not fit at all into Werner's system or do so only by means of very risky hypotheses, but I wish to confine myself to only the metal-ammonia salts to be discussed here. I must mention these salts as briefly as possible in order to give an idea, on the one hand, of how numerous such cases are and, on the other hand, of the impossibility of fitting them into Werner's system even with such risky hypotheses as assuming that $(H_2O)_2$ can act "bimolecularly" or that SO_4,H_2O can act as SO_4. In the following summary the numbers added in parentheses denote the water molecules which escape at 100°; those which are marked with an asterisk denote those which escape even on standing over sulfuric acid.[195]

1. *Hexammine Salts*

$$Ma_6 = R$$

RCl_3, H_2O M = Cr (1*), Rh (1*)

$R_2{\genfrac{}{}{0pt}{}{Cl_4}{PtCl_6}}, 2H_2O$ Co (1), Cr (1)

$R_2{\genfrac{}{}{0pt}{}{Cl_2}{(PtCl_6)_2}}, H_2O$ Co (0), Rh (0)

---

[192] *Ibid.*, *1*, 139 ff. [126–143 (1892)].

[193] [This has been shown to be $[Cu(H_2O)_4]SO_4 \cdot H_2O$.—G.B.K.]

[194] [This has been shown to be $[Fe(H_2O)_6]SO_4 \cdot H_2O$.—G.B.K.]

[195] a = NH_3, h = OH_2, en = ethylenediamine, p = pyridine.

Hexammine Salts Cont.

$R_2(PtCl_6)_3$, $6H_2O$[196]	Co (3), Co (3), Rh (3)	$R(PO_4)$, $4H_2O$	Co (4), Cr (4*), Rh (4*)
$R(HgCl_3)_3$, H_2O	Co (0*), Cr (0*)	$R_2(PO_4H)_3$, $4H_2O$	Co (3)
$R^{Br}_{(PtBr_6)}$, $2H_2O$[197]	Co (1)	$R_4(P_2O_7)_3$, $20H_2O$	Co (15*)
$R_2(PtBr_6)_3$, $4H_2O$	Cr (4)	$R_2(P_2O_7Na)_2$, $23H_2O$[198]	Co (22), Cr (21*), Rh (22)
$R_2(SO_4)_3$, $5H_2O$	Co (4), Cr (4*), Rh (4)	$R_2(C_2O_4)_3$, $4H_2O$	Co, Cr (0*)

M en_3 = R

RCl_3, $3H_2O$	M = Co (3)	$R_2(PtCl_6)_3$, $12H_2O$	Co(12)

2. *Aquopentammine Salts*

Ma_5h = R

$R_2{}^{Cl_4}_{(PtCl_6)}$, $2H_2O$	M = Co (3)	$R_2(SO_4)_3$, $3H_2O$	Co (4), Cr (4), Rh (4½)
$R_2{}^{Cl_2}_{(PtCl_6)_2}$, H_2O[199]	Co (3)	$R^{(NO_3)}_{PtCl_6}$, H_2O	Co (1), Rh (1)
$R_2(PtCl_6)_3$, $6H_2O$[200]	Co (5)	$R^{OH}_{PO_4H}$, H_2O	Co (2)
$R(HgCl_3)_3$, H_2O[201]	Co (2), Cr	$R_2(PO_4H)_3$, $4H_2O$	Co (5), Rh (5)
$R_2{}^{Br_2}_{(PtBr_6)_2}$, $2H_2O$	Co (4), Cr (4)	$R_4(P_2O_7)_3$, $12H_2O$	Co (6*)
$R_2(PtBr_6)_3$, $4H_2O$	Co (4)	$R_2(P_2O_7Na)_2$, $23H_2O$	Co (23), Rh (23)
$R^{OH}_{(S_2O_6)}$, H_2O	Co (2), Cr (2)	$R_2(C_2O_4)_3$, $4H_2O$	Co (0*)
$R_2(SO_3)_3$, H_2O	Co (1 at 80°)	$R_2{}^{(C_2O_4)}_{(PtCl_6)_2}$, $6H_2O$	Co (6)

3. *Diaquotetrammine Salts*

Coa_4h_2 = R

$R_2(SO_4)_3$, $3H_2O$ (3*, 3½)	$R_4(P_2O_7)_3$, $6H_2O$ (4*, 6)

4. *Iodopentammine Salts*

J·Rha_5 = R

$R(SO_4)$, $3H_2O$ (3*)

---

196 [The original erroneously reads "$R_2(PtCl_6)_2$, $6H_2O$."—G.B.K.]

197 [The original erroneously reads "$R_2{}^{Br_2}_{(PtBr_2)_2}$, $2H_2O$."—G.B.K.]

198 [The original erroneously reads "$R_2(P_2O_7Na)$, $23H_2O$."—G.B.K.]

199 [The original erroneously reads "$R_2{}^{Cl_2}_{(PtCl_6)}$, H_2O."—G.B.K.]

200 [The original erroneously reads "$R_2(PtCl_6)_2$, $6H_2O$."—G.B.K.]

201 [The original erroneously reads "$R(HgCl_3)_2$, H_2O."—G.B.K.]

5. *Bromopentammine Salts*

$BrMa_5 = R$

$R(SO_4), H_2O$ M = Ir (1*) $R(SO_4), 3H_2O$ Rh (3)

6. *Chloropentammine Salts*

$Cl \cdot Ma_5 = R$

$R(S_2O_6), H_2O$ M = Ir (1)

$R(SO_4), 2H_2O$ Co (2*), Cr (2*), Rh (2*), Ir (2*)

$R(CO_3), H_2O$ Co (1), Rh (0)

$R(CO_3), 9H_2O$ Co (9*)

$R(C_6H_5O_6)_2, 5H_2O$ Co (5)

$Cl \cdot Coen_2a = R$

$RCl_2, 2H_2O$ (2)

$R(PtCl_6), H_2O$ (1)

7. *Nitratopentammine Salts*

$NO_3 \cdot Ma_5 = R$

$R(S_2O_6), H_2O$ M = Co (1), Rh (1)

$R(SO_4), H_2O$ Co (1 at 135°)

$R(Cr_2O_7), H_2O$ Co (1*)

$R_3[Co(NO_2)_6]_2, 2H_2O$ Co (2*)

8. *Nitropentammine Salts*

$NO_2 \cdot Ma_5 = R$

$R(SO_4), H_2O$ M = Cr (1)

$R(S_2O_6), H_2O$ Cr (1), Rh (1)

$R_2(FeCy_6), 6H_2O$ Co (6*)

9. *Sulfatopentammine Salts*

$(SO_4 \cdot Coa_5) = R$

$R(SO_4H), 2H_2O$ (2)

$R_2(SO_4), H_2O$ (0)

10. *Oxalopentammine Salts*

$C_2O_4 \cdot Coa_5 = R$

$R(SO_4H), H_2O$ (1)

$R(SO_4H), 2H_2O$ (2)

$R_2(SO_4), 3H_2O$ (3*)

$2RJ, 3H_2O$ (3*)

$2RBr, 3H_2O$ (3*)

$R_2(PtCl_6), 2H_2O$ (0*)

11. *Sulfitopentammine Salts*

$SO_3 \cdot Coa_5 = R$

$R_2(SO_3), 2H_2O$

$R_2(SO_3), 2Na_2SO_3, H_2O$

12. *Sulfitoaquotetrammine Salts*

$SO_3 \cdot Coa_4h = R$

$R \cdot SO_3 \cdot NH_4, 2H_2O$

$R(NC), H_2O$

$R(CNS), H_2O$

13. *Dibromotetrammine Salts*

$Br_2 \cdot Coen_2 = R$ (Praseo)

$R_2(PtCl_6), 3H_2O$ (3)

14. *Dichlorotetrammine Salts*

$Cl_2Coa_4 = R$

$Cl_2Coen_2 = R$ (Violeo)

RCl, H_2O (1)

$2R(NO_2), 3H_2O$

RCl, H_2O (1)

15. *Carbonatotetrammine Salts*

$CO_3 \cdot Coa_4 = R$

$2R(AuCl_4), H_2O$

$R_2(PtCl_6), 2H_2O$ (1*, 2)

$R_2(SO_4), 3H_2O$ (3*)

$2R(NO_3), H_2O$ (1)

$R_2(CO_3), 3H_2O$ (3*)

16. *Sulfitotetrammine Salts*

$SO_3 \cdot Coa_4 = R$

$R \cdot SO_3 \cdot NH_4, 3H_2O$

$R \cdot SO_3 \cdot Na, 2H_2O$

$R \cdot SO_3 \cdot Rb, 2H_2O$

17. *Oxalotetrammine Salts*

$C_2O_4 \cdot Coa_4 = R$

$2R_2(PtCl_4), 3H_2O$ (3*)

$R_2(PtCl_6), H_2O$ (0)

$R_2(SO_4), 2H_2O$ (2*)

18. *Dinitrotriammine Salts*

$(NO_2)_2 \cdot Coa_3 = R$

$R_2(SO_4), 2H_2O$ (0*)

19. *Oxalotriammine Chloride*

$2C_2O_4(Coa_3)Cl, H_2O$ (0)

20. *Dinitrodiammine Oxalates*

$(NO_2)_2 \cdot Coa_2 \cdot C_2O_4 = R$

$R \cdot (NH_4), H_2O$ (1*)

$R \cdot K, H_2O$ (0*, 1)

$R \cdot Ag, H_2O$ (1*)

$R \cdot Na, 2H_2O$ (0*, 2)

$R_2 \cdot Ba, 3H_2O$ (3)

$R_3 \cdot Mg(NH_4), 6H_2O$ (0)

$R_3 \cdot Co(NH_4), 6H_2O$ (0*)

$R_3 \cdot Co(NH_4), 9H_2O$ (3*)

$R_3(Coa_6), 6H_2O$ (3*)

$R_3 \cdot CoAg, 9H_2O$ (3*)

In order not to lengthen this list unnecessarily, I have omitted complex [polynuclear] metal-ammonia salts, among which quite a few are hydrated.

As long as one is satisfied with general considerations of the close relationship between ammonia- and water-containing salts, or of the derivability of hydrated salts from higher types, or of the quite natural assumption that the individual components of higher molecular systems are arranged in the simplest three-dimensional or two-dimensional figures, much may seem conceivable at first sight. But here the most varied water-containing salts are listed in an extensively investigated group of metal-ammonia salts, and when confronted with this series, Werner's theory, with its claim of being able to explain these, is completely powerless since as far as I see it, a new hypothesis is necessary for almost every salt. I believe that it is better temporarily to leave with the older theory a task, to which our present knowledge is not yet equal, than to discuss all sorts of possibilities, of which the only certain thing is that they are just mere possibilities.

As far as the metal-ammonia salts are concerned, I have now compared the older theory with Werner's system from all essential points of view and as objectively as I could, and I am now even more convinced than previously that Blomstrand's theory together with my modifications still best explains all actual relationships in the chemistry of the metal-ammonia salts in accordance with our whole chemical system. It was my task to examine this question exactly as it relates to Werner's theory. For he is the only one who has tried to build a new theory considering all details. He has not been satisfied with casual suggestions, which are of absolutely no use if one wishes to apply them to the compounds discussed in this and previous work. Furthermore, it is to his credit that not only has he stimulated new researches by his many striking and ingenious observations, but he himself has revealed many new, interesting, and important facts, and I hope that he will reveal many more new facts.

The weak point of his theory is the octahedral formula. If it is compared with the ammonium radical of my formulas:

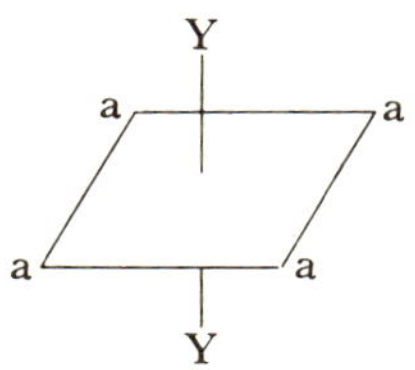

$$\begin{matrix} & \cdot Y \\ Co\cdot a\cdot a\cdot a\cdot a & \\ & \cdot Y \end{matrix}$$

they look very much alike. In both, Y can be replaced by NH_3, OH_2, or an electronegative radical (Cl, NO_2, NO_3, etc.). Also, in both formulas, 1 or 2 of the 4a can be replaced by OH_2. On closer consideration, however, it is found that almost all the weaknesses of Werner's theory that have been pointed out in this paper can be traced back to the fact that he assumes that the 6 positions in his octahedron are equal, whereas the older theory derives its strength from exactly the fact that it does not do this but on the contrary retains a definite difference between one-, two-, three-, and four-membered ammonia groups. Now, although the octahedral formula originally was without doubt the starting point for Werner's entire theory of inorganic compounds, it does not seem to be a necessary consequence of his theory as he has later developed it. Furthermore, the octahedral formula seems to have a rather accidental or arbitrary character, and besides, it seems to contain a contradiction within itself. If 1 atom of chlorine or another electronegative radical, for example, is thought of as being coordinated to the trivalent cobalt atom and does not saturate a valence of this atom, then it is incomprehensible why the radical $Co \cdot a_5Cl$ formed in this manner is only divalent, combines with only two electronegative ions, and does not function trivalently just as well as $Co \cdot a_6$ does.[202] On the other hand, if the chlorine atom saturates a valence of the cobalt atom, then it is not coordinated because coordinated atom groups do not of course change the valence of the metal atom.[202] In this case, one can certainly understand that the radical $CoCl \cdot a_5$ is only divalent, but now no octahedron can be formed because in this case only 5 coordinated atom groups are present. In any case, the octahedral formula is not necessary for Werner's system, as I now wish to demonstrate, just by drawing conclusions from his own considerations in the article on oxy salts and molecular compounds.[203] Here he regards sal ammoniac[204] as a coordinated compound in which the electronegative ion is not bonded directly to nitrogen:

$$\begin{array}{c} \text{H} \\ \cdot \\ \text{H} \cdot \text{N} \cdot \text{H—Cl.} \\ \cdot \\ \text{H} \end{array}$$

In the same paper he views pyrosulfuric acid and potassium dichromate as

---

[202] [From his first article on the coordination theory (*Z. anorg. Chem.*, *3*, 267–330 (1893); Kauffman, *Classics*, pp. 9–88), Werner recognized that regardless of whether an electronegative radical is bonded directly to the central metal ion (in the inner sphere) or indirectly (in the outer sphere), it reduces the charge on the central metal atom by an amount equal to its own charge.—G.B.K.]

[203] [A. Werner,] *Z. anorg. Chem.*, *9*, 382[–417 (1895)].

[204] [Ammonium chloride, NH_4Cl.—G.B.K.]

$$\begin{array}{ccccccccc} & & \mathrm{O} & & & & \mathrm{O} & & \\ \mathrm{O} & \cdot & \mathrm{S} & \cdot & \mathrm{O} & \cdot & \mathrm{S} & \cdot & \mathrm{OH_2} \\ & & \mathrm{O} & & & & \mathrm{O} & & \end{array} \quad \text{and} \quad \begin{array}{ccccccccc} & & \mathrm{O} & & & & \mathrm{O} & & \\ \mathrm{O} & \cdot & \mathrm{Cr} & \cdot & \mathrm{O} & \cdot & \mathrm{Cr} & \cdot & \mathrm{OK_2}, \\ & & \mathrm{O} & & & & \mathrm{O} & & \end{array}$$

and it can safely be thought that he also views Mitscherlich's potassium trichromate[205] and Siewert's potassium tetrachromate[206] in the same manner, i.e., as

$$\begin{array}{ccccccccccccc} & & \mathrm{O} & & & & \mathrm{O} & & & & \mathrm{O} & & \\ \mathrm{O} & \cdot & \mathrm{Cr} & \cdot & \mathrm{O} & \cdot & \mathrm{Cr} & \cdot & \mathrm{O} & \cdot & \mathrm{Cr} & \cdot & \mathrm{OK_2} \\ & & \mathrm{O} & & & & \mathrm{O} & & & & \mathrm{O} & & \end{array} \quad \text{and} \quad \begin{array}{ccccccccccccccccc} & & \mathrm{O} & & & & \mathrm{O} & & & & \mathrm{O} & & & & \mathrm{O} & & \\ \mathrm{O} & \cdot & \mathrm{Cr} & \cdot & \mathrm{O} & \cdot & \mathrm{Cr} & \cdot & \mathrm{O} & \cdot & \mathrm{Cr} & \cdot & \mathrm{O} & \cdot & \mathrm{Cr} & \cdot & \mathrm{OK_2}. \\ & & \mathrm{O} & & & & \mathrm{O} & & & & \mathrm{O} & & & & \mathrm{O} & & \end{array}$$

We have here a conception which in many ways recalls Berzelius' formulas, $2SO_3 + H_2O$, $2CrO_3 + K_2O$, $3CrO_3 + K_2O$, and $4CrO_3 + K_2O$.[207] But the path from Berzelius to Blomstrand is usually not long. In accordance with the above conception of the polychromate salts, it may be assumed that Werner too views the compounds $3NH_3,HNO_3$ of [B. B.] Kuriloff[208] and $4NH_3,HBr$ of [L. J.] Troost and [H. W.] Bakhuis-Roozeboom in the same manner, i.e.:

$$\begin{array}{ccccccccccccc} & & \mathrm{H} & & & & \mathrm{H} & & & & \mathrm{H} & & \\ \mathrm{H} & \cdot & \mathrm{N} & \cdot & \mathrm{H} & \cdot & \mathrm{N} & \cdot & \mathrm{H} & \cdot & \mathrm{N} & \cdot & \mathrm{H{-}NO_3} \\ & & \mathrm{H} & & & & \mathrm{H} & & & & \mathrm{H} & & \end{array} \quad \text{and} \quad \begin{array}{ccccccccccccccccc} & & \mathrm{H} & & & & \mathrm{H} & & & & \mathrm{H} & & & & \mathrm{H} & & \\ \mathrm{H} & \cdot & \mathrm{N} & \cdot & \mathrm{H} & \cdot & \mathrm{N} & \cdot & \mathrm{H} & \cdot & \mathrm{N} & \cdot & \mathrm{H} & \cdot & \mathrm{N} & \cdot & \mathrm{H{-}Br}. \\ & & \mathrm{H} & & & & \mathrm{H} & & & & \mathrm{H} & & & & \mathrm{H} & & \end{array}$$

It would be consistent with this to view, e.g., praseocobalt chloride[209] as

$$\begin{array}{ccccccccccccc} & & \mathrm{H} & & & & \mathrm{H} & & & & \mathrm{H} & & & & \mathrm{H} \\ \mathrm{H} & \cdot & \mathrm{N} & \cdot & \mathrm{H} & \cdot & \mathrm{N} & \cdot & \mathrm{H} & \cdot & \mathrm{N} & \cdot & \mathrm{H} & \cdot & \mathrm{N{-}CoCl_3}, \\ & & \mathrm{H} & & & & \mathrm{H} & & & & \mathrm{H} & & & & \mathrm{H} \end{array}$$

thus recalling [C.] Claus'[210] view of metal-ammonia salts. However, since in this salt only 1 chlorine atom is an ion and the arrangement naturally does not matter, then praseo chloride, always in accordance with Werner's above interpretation of di- and trichromic acids, can be viewed as

---

[205] [$K_2Cr_3O_{10}$ (E. Mitscherlich, *Lehrbuch der Chemie*, E. S. Mittler, Berlin, 4th ed., 1844, Vol. II, p. 759).—G.B.K.]

[206] [$K_2Cr_4O_{13}$ (M. Siewert, *Z. Ges. Naturwiss.*, (*Halle*), *19*, 11–33 (1862); *Jahresbericht*, 146 (1862)).—G.B.K.]

[207] [For a discussion of Berzelius' dualistic formulas see the Introduction to Paper Four in the present volume.—G.B.K.]

[208] *Z. physik. Chem.*, *25*, 108 [107–111 (1898)].

[209] [*trans*-$[Co(NH_3)_4Cl_2]Cl$.—G.B.K.]

[210] [See Paper Four of this volume.—G.B.K.]

$$\mathrm{Cl_2Co}\cdot\underset{\mathrm{H}}{\overset{\mathrm{H}}{\mathrm{N}}}\cdot\mathrm{H}\cdot\underset{\mathrm{H}}{\overset{\mathrm{H}}{\mathrm{N}}}\cdot\mathrm{H}\cdot\underset{\mathrm{H}}{\overset{\mathrm{H}}{\mathrm{N}}}\cdot\mathrm{H}\cdot\underset{\mathrm{H}}{\overset{\mathrm{H}}{\mathrm{N}}}\cdot\mathrm{H}\text{—}\mathrm{Cl},$$

an assumption which in many ways recalls my formula for praseo chloride:

$$\mathrm{Cl_2\cdot Co\cdot NH_3\cdot NH_3\cdot NH_3\cdot NH_3\cdot Cl}$$

but has nothing at all in common with the octahedral formula. Perhaps Werner's and my theories will thus merge into a higher unity.

Copenhagen, Laboratory of the Polytechnische Lehranstalt [*Polytekniske Laereanstalt*], *September 1898.*

Received by the editor on Sept. 22, 1898.

Supplementary Reading

Bailar, J. C., Jr., "The Early Development of the Coordination Theory," in *The Chemistry of the Coordination Compounds*, J. C. Bailar, Jr., ed., Reinhold Publishing Corp., New York, 1956, Chap. 2. A short chemical-historical treatment of theories of complex compounds through the time of Werner (19 pp.).

Basolo, F., and Johnson, R., *Coordination Chemistry*, W. A. Benjamin, New York, 1964. A short introduction intended as supplementary reading for general chemistry students (180 pp.).

Graddon, D. P., *An Introduction to Co-ordination Chemistry*, Pergamon Press, New York, 1961. A short but scholarly introduction to the subject (111 pp.).

Grinberg, A. A., *An Introduction to the Chemistry of Complex Compounds*, transl. by J. R. Leach; D. H. Busch and R. F. Trimble, Jr., eds., Pergamon Press, New York, 1962. A comprehensive treatment emphasizing history and experimental data, especially the work of Russian chemists (363 pp.).

Jones, M. M., *Elementary Coordination Chemistry*, Prentice-Hall, Inc., Englewood Cliffs, N.J., 1964. A comprehensive introduction for chemists and workers in other fields (473 pp.).

Kauffman, G. B., *Alfred Werner, Founder of Coordination Chemistry*, Springer-Verlag, Berlin, Heidelberg, New York, 1966. A nontechnical biography (127 pp.).

Kauffman, G. B., *Classics in Coordination Chemistry, Part I: The Selected Papers of Alfred Werner*, Dover Publications, Inc., New York, 1968. This book contains annotated translations of Werner's six most important articles, each with an introductory essay and commentary (190 pp.).

Kauffman, G. B., *Classics in Coordination Chemistry, Part III: Selected Papers, (1904–1935)*, Dover Publications, Inc., New York (in preparation). Annotated translations or reprints of ten articles by Heinrich Ley, Lev Aleksandrovich Chugaev, John Albert Newton Friend, Paul Pfeiffer, Ralph W. G. Wyckoff and Eugen Posnjak, Roscoe G. Dickinson, Nevil Vincent Sidgwick, Il'ya Il'ich Chernyaev, and William H. Mills and Thomas H. H. Quibell, each with an introductory essay and commentary.

Lewis, J., and Wilkins, R. G., *Modern Coordination Chemistry: Principles and Methods*, Interscience Publishers, New York, 1960. Six essays on thermodynamics, kinetics, isomerism, spectra, and magnetochemistry (487 pp.).

Martin, D. F., and Martin, B. B., *Coordination Compounds*, McGraw-Hill Book Co., Inc., New York, 1964. A short, elementary introduction at the beginning undergraduate level (99 pp.).

Murmann, R. K., *Inorganic Complex Compounds*, Reinhold Publishing Corp., New York, 1964. A short monograph intended as supplementary reading for general chemistry students (120 pp.).

Orgel, L. E., *An Introduction to Transition-Metal Chemistry: Ligand-Field Theory*, 2nd ed., John Wiley and Sons, New York, 1966. A brief treatment of this important theory (186 pp.).

Quagliano, J. V., and Vallarino, L. M., *Coordination Chemistry*, D. C. Heath and Co., Lexington, Mass., 1969. A condensed treatment for general chemistry students. Includes exercises and problems (124 pp.).

Read, J., *Humour and Humanism in Chemistry*, G. Bell and Sons, London, 1947, pp. 262–284. Read's personal reminiscences of his stay in Zürich as one of Werner's *Doktoranden*.

Index